Cambridge County Geographies

GENERAL EDITOR: W. MURISON, M.A.

BERWICKSHIRE
&
ROXBURGHSHIRE

Cambridge County Geographies

BERWICKSHIRE
&
ROXBURGHSHIRE

by

W. S. CROCKETT

Minister of Tweedsmuir
Member of the Berwickshire Naturalists' Club
Author of *The Scott Country*, *The Scott Originals*, etc.

With Maps, Diagrams and Illustrations

CAMBRIDGE
AT THE UNIVERSITY PRESS
1926

CAMBRIDGE UNIVERSITY PRESS
Cambridge, New York, Melbourne, Madrid, Cape Town,
Singapore, São Paulo, Delhi, Mexico City

Cambridge University Press
The Edinburgh Building, Cambridge CB2 8RU, UK

Published in the United States of America by Cambridge University Press, New York

www.cambridge.org
Information on this title: www.cambridge.org/9781107623873

© Cambridge University Press 1926

First published 1926
First paperback edition 2013

A catalogue record for this publication is available from the British Library

ISBN 978-1-107-62387-3 Paperback

PREFATORY NOTE

THIS combination of the two Border Counties has been found to be more satisfactory than in treating each County separately. Much has necessarily been omitted on account of the restricted nature of the volume. For further enquiry there may be commended the exhaustive *History* of the Berwickshire Naturalists' Club, begun in 1831, and continued to date; the *Transactions* of the Hawick Archaeological Society, from 1856 onwards; and the Berwickshire Report and Inventory by the Royal Commission on the Ancient and Historical Monuments and Constructions of Scotland, published in 1915.

I have to express indebtedness to the following who gave help upon several points: Mr J. Hewat Craw, Secretary of the Berwickshire Naturalists' Club; Mr T. Cuthbert Day, F.G.S., who assisted with Chapter Six; the Rev. Alexander Grieve, M.A., D.Phil., who put at my disposal (for Chapter Ten) his knowledge of Border Philology; and to Mr James Logan Mack for information as to the Border Line described in Chapter Three.

W. S. CROCKETT

The Manse,
  Tweedsmuir,
January 1926

CONTENTS

ILLUSTRATIONS

MAPS

The illustrations on pp. 12, 51, 53, 54, 77, 85, 86, 93, 111, 116, 118, 121, 130, 132, 133, 144 (lower), 183, 187, 188 are reproduced from photographs by Messrs J. Valentine and Sons, Ltd.; those on pp. 55, 119, 128, 135, 136, 144 (upper) from photographs by Miss Cameron, Duns; those on pp. 6, 21, 114, 134 from photographs by the late Rev. D. G. Manuel; those on pp. 24, 145 from photographs by Dr Blair; those on pp. 113, 139, 141, 146 from photographs by Miss McIntosh; that on p. 143 from a photograph by the Rev. Ae. E. M'Innes; those on p. 101 are reproduced by permission of the Berwickshire Naturalists' Club; that on p. 171 by permission of the Hawick Archaeological Society; those on pp. 97, 104, 109 by permission of the Society of Antiquaries of Scotland; those on pp. 107, 125 by permission of the Royal Commission on Ancient Monuments; that on p. 166 is reproduced from a photograph by Mr Alexander Ayton, Edinburgh; that on p. 177 is from a photograph supplied by Miss H. M. R. Murray; that on p. 158 is reproduced (by permission) from a sketch by A. E. Chalon, R.A., painted in 1839; that on p. 127 is from a private photograph; the diagrams are by Miss Ray Somerville, M.A., B.Sc. The map on p. 123 is by the author.

BERWICK & ROXBURGH

1. County and Shire. The Origin of Berwickshire and Roxburghshire.

A division of the country into counties and parishes appears to date from the time of David I, and it was a result of the feudal system then being applied to Scotland. The monarch parcelled out portions of the realm to favourite knights whom he created *counts*, a title borrowed from the French *comte*, from Latin *comes*, a companion, and equivalent to the older English term *earl*. The *comitatus*, or county, was thus a district under the jurisdiction of a king's *comes*.

The word *shire* comes from the Anglo-Saxon *scir*, office, administration, and meant at first an area marked off for ecclesiastical purposes, as in Coldinghamshire and Norhamshire. This corresponded no doubt to what afterwards was called the *parochia*, or parish. In course of time *shire* altered its meaning. Broadened and secularised, it came to denote the whole of a county over which was set the *scir-grefa*, shire-reeve, sheriff, an office not to be confounded with that of the modern sheriff. The sheriff was *vice-comes*, a mediæval Lord-lieutenant, but with more extensive powers, concerned not only with the administration of justice but responsible also for the military affairs of the shire.

Berwickshire is in the anomalous position of being a Scottish county bearing the name of an English town. In the course of its chequered story Berwick-upon-Tweed changed hands no fewer than thirteen times. In 1482 it was attached to the English Crown, and is now a county in itself. The name is derived either from the Anglo-Saxon *bere*, barley, and *wic*, a dwelling or village, i.e. the "demesne-farm," or from *bar*, bare, and *wic*, "the naked, unsheltered homestead." Less must be said for *bearh* and *wic*, "the castle town." Still less for the familiar *aber*, "at the mouth of," and *wic*—a form unrecorded in any early document. The story of the original settlers finding the country infested with "bears" and taking a "week" to slay them is a mere canting explanation, though the seal of the Burgh of Berwick (a specimen of the year 1250 is extant) exhibits a good-sized grizzly chained and collared beside a tree with the ocean in the background.

Berwickshire is sometimes styled the Merse. Strictly speaking this is only one of the three parts into which the county is divided.

Roxburghshire has its designation from the long-demolished Castle of Roxburgh overlooking the rivers Tweed and Teviot, or perhaps from the extinct and important burgh of Roxburgh upreared on the plain beneath. The word may mean simply "the castle on the rock" (rock's burgh), only "rock" is not found as an English word before Chaucer. Sir Herbert Maxwell suggests "Rawic's burgh" as a derivative, but of Rawic we know nothing. Twelfth-century spellings were Rokeburg, Rokesburh, Rokisburgh, and Rocheburch. A more ancient name, Marchidun, "the castle on the Marches," belongs to the British period.

From its principal stream, the Teviot, this county was known of old as Teviotdale, colloquially Tividale (so called in 1300) and Tibbidale. In a map of 1292, preserved in the British Museum, in which the castles of Roxburgh and Jedburgh are prominent, the whole of this part of the Border is included under the designation Tynedale (the Northumbrian Tyne), while Lothian embraces all Berwickshire—all between the Forth and the Tweed.

2. General Characteristics.

Berwickshire and Roxburghshire are pre-eminently the Border Counties. Though the whole valley of the Tweed is regarded as the Border from a literary and romantic point of view, neither Peeblesshire nor Selkirkshire touch anywhere the actual line of demarcation. On the other hand, Berwickshire has Northumberland for almost one-fourth of its southern boundary. Roxburghshire marches with Northumberland and a considerable portion of Cumberland. Here it was a quite natural barrier which divided the kingdoms of Scotland and England—the river Tweed and "Cheviot's mountains lone."

The counties lie in what is known as the Southern Uplands section—the third of the three parts into which Scotland is divided physically, the others being the Highlands, and the Middle or Rift Valley, whose southern delimitation may be taken to be a line running from Dunbar to Girvan Bay. All south of that line, therefore, to the Cheviots and the Solway constitutes the Southern Uplands. Nature has bestowed on our counties equally well-defined tripartite divisions. Between the Lammermoors and the Tweed stretches the broad sloping plain of the Merse, upwards of 120,000 acres in extent. Practically the whole of it has been put under cultivation—utilised with such skill in agriculture and its accompanying adornments as to gain the distinction of being the largest and richest tract of low country in Scotland. In the Lammermoor (probably from Celtic *lann barra mor*, "the level spot on the big height") series of hills and the fine pastoral country at their base, confined to the boundary-parishes of Westruther, Longformacus, Cranshaws, and Abbey St Bathans, we have the second physical division of the shire. The third is Lauderdale, the valley watered by the Leader and its tributaries.

In Teviotdale, the ancient designation of Roxburghshire, we have the largest and most important of its divisions, including not merely the basin of the Teviot, but also so much of the Tweed valley as is included within the shire. Part of Lauderdale (probably Celtic *liath dobhar*, "the grey stream")—its west-most portion, confined to the extensive parish of Melrose, is a second natural division. The other is Liddesdale, situated on a different watershed, and included in the basin of the Solway.

The characteristics of the counties are much alike—rolling stretches of green, undulating country, a clear smokeless atmosphere, ranges of broad, grassy hills to north and south, with several solitary heights and lesser eminences, many beautiful towns, villages and hamlets, ancient and modern churches, stately woodlands and smiling hedgerows lying between. Berwickshire is partly a maritime county, having the North Sea for its eastern boundary, while Roxburghshire is entirely inland. Mechanical industries are sparse except for the manufacturing interests of Hawick and some smaller towns. There is no mineral wealth, stone quarries are abundant, and the fisheries of Berwickshire are limited to a few creeks or stations. The chief industry of the counties is agriculture. Everywhere we pass the comfortable farm-house with its well-appointed steading and labourers' cottages, reputed to be the best in the land. Nowhere are the fields brought so thoroughly under cultivation, so little available ground being lost, so much reclaimed and gained from the moor and the marsh. Even the hill-foots of the Lammermoors and of the Cheviots provide their quota of pastures green, but waving cornfields we may see there also, though harvest is naturally later than in the vales beneath.

In point of picturesqueness, while the Border cannot claim the wild, often weird, grandeur of Highland scenery, it is ever an attractive landscape that presents itself, affording a constant charm to the eye and furnishing a rich field of inspiration for the painter's brush and the poet's pen. Elihu Burritt declared that one of the finest views in Scotland is that which gripped the soul of Scott as with a magnet when from the shoulder of Bemersyde Hill he gazed at the majestic panorama spreading out in front of him. There are many points of vantage all over Berwickshire and Roxburghshire for the lover of natural beauty. From the Eildon Hills, the Black Hill of Earlston, the Dunian and Ruberslaw, Smailholm Crags, Hume Castle, or the heights above Chirnside, we may revel in pictures of almost unmatched loveliness. And on a less expansive scale both counties abound in hundreds of peaceful nooks by riverside and sequestered valley, in all of which the pictorial element is striking and memorable. The Berwickshire coast is a

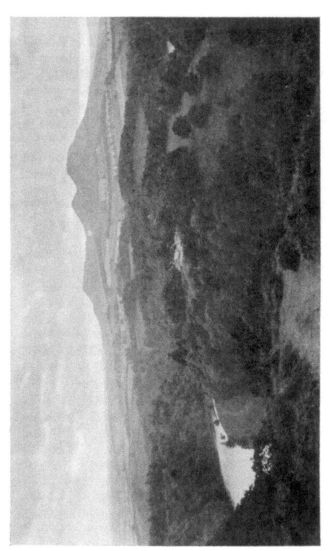

The Eildons and the Tweed
[Scott's favourite view from Bemersyde Hill]

happy hunting-ground for the geologist, the deans of both counties a botanist's paradise.

A leading feature of the Border district is the wealth and variety of its rivers, its streams, and mountain-burns. But for these our counties would not be what they are. The silver Tweed is in itself a realm of romance, for no Scottish stream—perhaps no stream in the world—has been so besung. Nor can less be said of its fair daughters, the Leader and the Teviot:

> "A mist of memory broods and floats,
>   The Border waters flow;
> The air is full of ballad notes
>   Borne out of long ago."

The Past is never far away. The shattered "peel" is a frequent landmark. The noblest monastic ruins in Scotland are here. Every mile we travel is on historic ground. Not a field but has its battle. Not a rivulet but has its song. Great and famous names are being constantly conjured up. Over all is the spell of that Old Romance and that later wizardry of the "King of the Romantics" which have carried the name of the Border to the furthest regions of the globe.

The history and traditions of a people are always influenced by the character of a land in which they dwell. So even the physical features of these shires, their place in the group of Scottish counties, their contours, the contribution they make in the general economy of the State, cannot be useless as explaining many important events connected with the fortunes of the nation, and illustrating numerous local allusions in its folk-lore and its poetry.

3. Size. Shape. Boundaries.

Of our two counties Berwickshire is the smaller, with an area of 292,535 acres, or 457 square miles, exclusive of inland water, tidal water and foreshore. The latter extends to about 1503 acres, or 35 square miles. Of the thirty-three Scottish counties Berwickshire is nineteenth in point of size. It is nine and a half times less than the largest—Inverness-shire—but it could contain Clackmannanshire—the smallest—more than eight times. Berwickshire and Roxburghshire are each bigger than their neighbours of the Border quartet, and combined, they are larger than Dumfriesshire, which ranks eighth in size among Scottish counties. Between them and Ayrshire there is only the difference of a good-sized pastoral farm.

Berwickshire is oblong in shape. Its greatest length from Lamberton Beach, on the east, to the extreme west of Channelkirk is approximately 34 miles. From north to south—Cranshaws to Coldstream—its utmost breadth is 22 miles. Exclusive of minor sinuosities its coast-line has a length of nearly 19 miles.

The county is bounded on the north by East Lothian, on the west by Midlothian, on the south by Roxburghshire and Northumberland, and its eastern border is the North Sea. It is the most south-easterly county of Scotland For about 40 miles it is separated from Roxburghshire and Northumberland by the river Tweed.

The northern boundary begins at the mouth of the Dunglass Burn and follows that stream to its confluence with Oldhamstocks Burn. The tiny Berwick Burn now becomes the march, which is continued along the Dod Strip to the top of Dod Hill (1147 feet). Descending to south-west, the Eye Water is crossed, where the line turns obliquely south-east, then southward, and again south-west to the Monynut Water. It then follows the Philip Burn and comes down to the Whitadder, where a general westward trend begins. Climbing the watershed between the Dye and the Fasney streams—both tributaries of the Whitadder—it passes over the summit of Meikle Says Law (1750 feet), the highest of the Lammermoors. Near Lammer Law it pursues the water parting between the Haddingtonshire Tyne and the Leader, and proceeding south by Ninecairn Edge (1476 feet) and Soutra it unites with the Armet Water, which delimits the whole of the western portion of Channelkirk. Thence it cuts in by Muircleuch to St Leonards and Bridgehaugh, where it joins the Leader, which now becomes the boundary for about 18 miles, to its junction with the Tweed at Drygrange. Tweed is then the march to the extreme south-east portion of Mertoun, but here, breaking away from what should have been the natural boundary, the line turns directly northward again, until, winding its tortuous way round the eastern portion of Earlston parish, it coalesces with the Eden, abuts on the policies of Floors Castle, and then bending back to west and north and east by Mainberry, Sweethope Crags, and Hassington Mains, it finally turns south to link itself with the Tweed once more.

These boundary-lines are of the most capricious character. Geographical features scarcely determine them at all. The apparently natural march line—that of the rivers and streams—is only occasionally followed. So with the

more or less well-defined ridges of the hills. Not infrequently the boundary dovetails to the limits of an estate
or of a farm, though perhaps the estate or the farm has
not been able to push itself further because of those ancient
and well-nigh sacred marches that have remained the same
for centuries.

The Boundary Commissioners of 1889 rectified some
of these old arrangements. A portion of Oldhamstocks
which had lain in Berwickshire was transferred to East
Lothian in exchange for a detached part of Oldhamstocks
(Butterdean) in Coldingham parish. At Earlston a detached part at Craigsford was transferred to Melrose. A
similar transference was made at Kedslie in Lauder parish,
and a small portion of Mertoun was given to St Boswells.
Other parish boundaries at Abbey St Bathans, Cockburnspath, Cranshaws, Longformacus, Hume, Nenthorn and
Eyemouth were adjusted.

Among Scottish counties Roxburghshire is fourteenth
in order of size. It is thirteen times as large as Clackmannanshire, and one-sixth the size of Inverness-shire. It
is barely one and a half times as large as Berwickshire.
Together, Berwickshire and Roxburghshire are three and
three-quarter times smaller than Inverness-shire, and they
are about one and one-sixth times the size of Peeblesshire
and Selkirkshire. The land acreage of the county is
426,028 acres, and the water acreage, 2800—a total of
428,828 acres, slightly less than 666 square miles.

Its greatest length, from south-west to south-east, is about
44 miles, and it is 28 miles in the transverse direction.
Exclusive of the northern part of Melrose parish, extending
to about 25 square miles, the county constitutes an irregular

rhomboid, of which the south-east side from Liddelbank to Ardhope Cairn measures 38 miles, the north, from the latter point to Stichill, 28 miles; the north-west side from Stichill to Moodlaw Loch 35 miles, and the remaining or south-west side—from Moodlaw to Liddelbank, 21 miles.

A walk along an imaginary straight boundary-line would thus cover 122 miles, but the irregularities are innumerable, and the whole walk, were it possible, would be not less than 150 miles.

Berwickshire bounds the county on the north: on the east and south-east it is met by Northumberland: on the south-west by Dumfriesshire: on the west by Selkirkshire, and on the north-west by a south-eastern projection of Midlothian.

Starting at a point where Tweed leaves the county at Carham, we follow that river for a short distance in a semicircular direction. Thence moving to the north-west and skirting the parishes of Eccles and Stichill, the line proceeds almost due west to Sweethope, where it begins to find its way south again and bends suddenly back in curiously fantastic curves by Newton-Don and Edenbank where the Eden takes up the tale, carrying the boundary westward again by Easter Muirdean and Nenthorn House to a point near Mellerstain. Passing Rachelfield and Covehouse, it now turns in a south-easterly direction by Smailholm Crags till it meets the Tweed a few hundred yards from Rutherford Lodge. The Tweed is now the boundary to Leaderfoot, and the Leader the boundary to about Nether Blainslie. Almost within sight of Lauder the north-most boundary is reached. Soon we descend the watershed between the Leader and the Gala. Near Bowland we touch three counties—Roxburghshire, Selkirkshire, and

Hen Hole
Cheviot Hills. Boundary between Scotland
and England

Midlothian. Following the Gala to the south-east as far as Galafoot, then west again by Abbotsford, and again south by Cauldshiels Loch, an almost due southerly direction is traced by Whitelaw (1059 feet) to Moorfield, where the line turns westward to the Ale Water boundary between Ashkirk and Roberton. Northwards it turns once more to the Craigs (1238 feet) and again south by Hellmuir Loch and Kingside Loch to the Rankle Burn and Moodlaw Loch, the most westerly part of the county, where the Dumfriesshire border is reached. From here onwards to Whitehope Edge (1560 feet) it passes in a south-easterly direction, turns then to the east by Causeway Grain (1607 feet) and Wisp (1950 feet) crossing the Hawick and Langholm road at Mosspaul, and running up by Tudhope Hill (1961 feet) resumes a southerly course to Pike Fell (1637 feet). For the last time it goes to the east—to Hartsgarth Fell (1806 feet), and for the last time south—to Roan Fell, and hugging the watershed between Ewes and Liddel continues its course to Tinnis Hill and its south-most boundary at Liddelbank, where it abuts on Cumberland and Dumfriesshire. The Liddel for 6 miles is the boundary to Kershopefoot, the Kershope being the boundary for other 9 miles. At Lamisik Ford three counties—Roxburghshire, Cumberland and Northumberland—again meet. Farther on the line strikes north by east to Lariston Fell (1677 feet) and Buckside Knowe. Passing Bell's Burn it works its way to the top of Thorlieshope Pike (1180 feet), where again it turns east to the Marsh of Deadwater. Soon it crosses the railway near Deadwater station and proceeds to the highest point of Peel Fell (1975 feet). Crossing between the Scottish and English cairns on the summit it makes for Keilder Stone and Wylies Craigs, then falls down to the confluence of the Green and Black Needles. By and by it climbs Carter Fell (1815 feet) and passes over Catcleuch Shin (1742 feet) to the main road at Carter Bar (1250 feet) whence it continues its course by Arks Edge, Leap Fell (1542 feet), Fairwood Fell, to the top of Hungry Law (1643 feet). Turning eastward it makes for Greyhound Law and the Hearts to Coquet Head near the Roman Camp at Chew Green. At Brownhartlaw (1664 feet), it crosses the track of the old Roman road. Next it makes for the tops of Black Halls, Black Hall Hill (1573 feet) and Rushy Fell (1580 feet). From there it follows the ridges of Lamb Hill (1676 feet), Beefstand Hill (1844 feet) and Mozie Law (1812 feet) to Windygyle (2034 feet) and Lord Russell's Cairn. Crossing the old road from Cocklawfoot to Uswayford at Hexpethgatehead, it continues on its way to a small grassy mound near the "Hanging Stone on Cheviot." At this spot the line reaches its greatest elevation (2419 feet). It runs northward now to the top of Auchhope Cairn (2382 feet), over Auchope Rig to the top of the Schel (Sowerhopeshiel, 1979 feet) and down its northern slope to the Black Hag (1790 feet) and the Halterburn. Near Yetholm Mains it reaches the Bowmont Water. Crossing that stream and the Yetholm and Mindrum road, it follows the Reddenburn to a point below Pressenmill. Carham Burn takes it to the Tweed.

Under the Local Government (Scotland) Act, 1889, the parish of Ashkirk was placed wholly in Selkirkshire. The old parish of Lindean, conjoined with Galashiels but situated in Roxburghshire, was handed over to Selkirkshire. Roberton, partly in Selkirkshire and Roxburghshire, was placed entirely in the latter county. The parish of Selkirk, partly in Selkirkshire and Roxburghshire, was placed wholly in Selkirkshire, and a detached portion at Todrig was transferred to Ashkirk. Adjustments of anomalies were also made in Hawick, Wilton, Jedburgh, Oxnam, and Southdean.

4. Surface and General Features.

The triple division of Berwickshire is indicative of its physical conditions. Lammermoor, lying mainly to east and north, is its upland region—an extensive curvature of mostly bare and cheerless heights, nowhere bold and imposing, and often subsiding into low rolling tablelands of bleak moor. The range is really a continuation of the vaster heights on the west which have their beginning at Cheviot itself. But for the cleft made by the vale of the Gala, the Lammermoors, Peeblesshire, Dumfriesshire, and Lanarkshire (or Lowther) Hills, would be one unbroken chain. All are parts of that great geological fault—that plateau of ancient rocks—which runs across the country, and, broadly speaking, differentiates Highland from Lowland Scotland.

No Lammermoor hill-top has the loftiness of its Peeblesshire neighbours, but nearly all are exceeded by heights within the Cheviot range. Lammer Law (in East Lothian) is 1733 feet above sea-level. Seenes Law (in Lauder parish) comes next with 1683 feet, followed by Crib Law (1670), Willie's Law (1626), Hunt Law (1625), Wedderlairs (1593), North Hart Law (1578), Meikle Law (1531), Waddels Cairn (1490), Ninecairn Edge (1479), Hog's Law (1470), Wedder Law (1462), South

Hart Law (1437), Hog Hill (1395), Wether Law (1379), Lamb Rig (1339), Byrecleugh Ridge (1335), Black Hill (1299), Dun Law (1292), Tarf Law (1248).

These all lie in the western or middle sections where the altitudes are greater than those of the east, which seldom rise beyond 800 or 900 feet. Grass-covered to the summit, the gradient is generally easy. Once at the top, the plateau-like character of the scene is apparent. It is startling to be reminded that millions of ages ago the height on which one may be standing and looking out to the Cheviots and the Tweed uplands in the dim distance, was once a vast ocean-bed over which waves lashed in furious foam and sea-birds shrieked and flew amid the war of waters.

As nearly all the Border area rests upon sandstone and silurian, its general aspect is of a softer and serener type than pertains to a region where the bedrock has been less liable to wear down. Such is the way in which the Lammermoors and the Cheviots have been carved out through countless epochs of geological time. Such, too, is how the garden-like Merse has been formed. The softer heights have been worn away, and with the rich detritus of boulder clay carried down by glaciers from the uplands we are inheritors of its enduring legacy.

Lammermoor, then, is the moorish district of Berwickshire, a considerable grazing expanse computed at 140 square miles, consisting of a broad range of rounded but lofty hills stretching from the extreme north-west to Fast Castle and St Abbs Head. It is intersected by numerous valleys, each with its own little burn fed by sykes and rills, filtering through neighbouring bogs and forming in its turn the sources of the Leader, the Dye, the Eye, the

Whitadder and the Blackadder. Within a hundred years the advance of cultivation has wonderfully altered the complexion of those desolate but healthy Berwickshire uplands. Thousands of acres of once waste, bare, swampy land have been reclaimed. Cattle and sheep roam over green and luscious pastures. Even the cornfield is present, but higher up the purple heather still scents the air. The Lammermoors, all lying within Berwickshire and East Lothian, constitute for nearly two-thirds of their extent a southern screen to the latter county, and they form the north-most boundary of that great triangular basin of the Tweed over which the statelier line of the Cheviots stands sentinel upon south and east.

Lauderdale gives name to the fair and fertile valley of the Leader, a romantic admixture of a hundred square miles of wild and successfully cultivated strath-land, in which are situated some of the finest pastoral and arable farms of the Lowlands.

The Merse has nothing to do with the word "March" in the sense of a boundary. It signifies "low land" such as usually margins a river, as when we speak of Leader "haughs." Interspersed between the Lammermoors and the Cheviots are several isolated heights, some of them volcanic in their origin, and each inscribed upon the historic page of Scotland: Duns Law (713 feet), the crags of Hume (700) and of Smailholm (680), Earlston Black Hill (1031), the triple Eildons (1385, 1327, 1266), the Dunian (1095), Penielheugh (774), Ruberslaw (1392), Minto Crags (729), Minto (905), and Dunlaw (683), until we reach the Liddel Water Hills and the bounds of Cheviot.

The Cheviot Hills extend from the head waters of the Liddel to Yeavering Bell and the heights around Wooler, in Northumberland, a distance of over 30 miles. The Great Cheviot (2676 feet) which gives its name to the range is wholly an English mountain, but Auchhope Cairn (2382 feet), Windygate Hill (2034), Hungry Law (1643), Carter Fell (1815) and Peel Fell (1975), all culminate on the direct Border line. Their general character is that of smooth rounded masses with long flowing outlines. There are no peaks, no serrated Highland ridges; and the valleys as a rule show no precipitous crags and rocky precipices. The hills fall away with a long gentle slope into England, while on the Scottish side the descent is somewhat abrupt. They present also a more picturesque appearance on the Scottish than on the southern side, where the flat-topped elevations are covered for the most part with peat.

On the Teviotdale and Liddesdale watershed the chief summits are Wheelrig Head (1465 feet), Carlintooth (1801), Needslaw (1457), Fan Hill (1643), Windburgh (1622), Leap Steil (1544), Maiden Paps (1677), and Skelfhill Fell (1749), while on the Dumfriesshire border are Tinnis Hill (1326), Black Edge (1461), Watch Hill (1642), Roan Fell (1862), Hartsgarth Fell (1806), Din Fell (1735), Tudhope Hill (1961), Wisp Hill (1950), Whitehope Edge (1560), Ewesdown Fell (1468), and Craik Cross (1482).

Fertile fields and delicious pasturclands have long since taken the place of the old natural forests of our shires. Occasional scraggy reminders occur here and there, and many place-names containing *wood, shaw, tree,* and *aik* (Harewood, Headshaw, Sorbietrees, Aikieside) testify to

their former abundance and vastness. Large oak-trunks
have been dug out of moors and mosses—contemporaries
perhaps of those which the aborigines hollowed out with
fire to shape their canoes.

5. Rivers and Lochs.

It is their rivers that form the chief natural feature of our
counties. Perthshire alone is a more be-channelled shire.
From the Tweed to its smallest tributary, and to their
tributaries, each has its own character and its own story.
Not one without its song. Like rims of a massive green
cup holding in its hollow the waters of Tweed and Teviot,
the Lammermoors and Cheviots constitute a unique
watershed for the innumerable burns and streams which
seek the greater rivers. Liddesdale, separated from he
main portion of Roxburghshire, sends its waters to he
Solway. In Berwickshire the Eye is the only considerable
stream that finds the sea.

The entire drainage basin of the Tweed covers an area
of about 1870 square miles. In this five-sixths of Berwickshire are comprehended and the whole of Roxburghshire,
except Liddesdale.

For 30 miles of its course the Tweed flows through
Roxburghshire. From its source at Tweed's Well (1500
feet) in the Peeblesshire uplands, it is a Border river for
96 miles before it quits the county and acquires that international character which has made its banks the emblems
of two nationalities. The river enters Roxburghshire from
Selkirkshire a short distance from Lindean near the influx

of the Ettrick at nearly 300 feet above sea level. After passing Faldonside, it fronts the towers of Abbotsford, the most celebrated mansion on its banks. Here Sir Walter Scott lived till 1832. Here he died. Below Abbotsford the river receives the waters of the Gala. A little further on, it passes Bridgend, so called from a bridge built by David I to afford a passage to Melrose Abbey. Part of it could be seen in 1746, but all is gone now. This is the country of the *Monastery*. Below Bridgend the river receives the rapid waters of the Allan, or Elwyn, after dancing their way through the Fairy Dean from the haunted pastoral vale in which are situated the three towers of the novel—Glendearg, the home of the Glendinnings, Langshaw, and Colmslie. Darnick is the quaintest of Border villages with its finely-preserved peel, a store of Border antiquity. The Waverley Hydropathic is within a short distance, and Melrose comes in sight with its "dark Abbaye," an object of marvellous beauty despite its dull and prosaic modern environment. The river now winds round old-world, sun-dialled Newstead, and passing with a noble sweep beneath the cleverly-engineered viaduct of the Berwickshire railway and the older road-bridge leading into Lauderdale, it assumes what many consider to be its acme of loveliness and sylvan grandeur. At Leaderfoot it is joined by the Leader. Many classic spots are in the vicinity—

> Drygrange with the milk-white yowes,
> Twixt Tweed and Leader standing,

Gledswood, Bemersyde, the stately horse-shoe bend at Old Melrose on whose peninsula stood the original Melrose (*mel-ros*, bare promontory) of which nothing is left to

proclaim its old-time sanctity and renown. Dryburgh is environed by another picturesque bend. Shortly after quitting
this hallowed spot Mertoun is passed—long the home of a
Scott. Makerstoun now comes into view, and soon we
arrive at ruined Roxburgh, once a veritable "key of
Scotland," to which palatial Floors offers a not unkindly
foil. At Kelso we witness what Andrew Scott has besung

The Rhymer's Tower, Earlston

as the "marriage" of Tweed and Teviot, where Tweed is
only 83 feet above the sea, and has a width of over 100
yards. Moving with kingly course by Pinnaclehill and
Hendersyde, the river receives the waters of the Eden.
Then follow Sprouston, Birgham, Coldstream, and farewell to the county is said at a point near Carham.

Kelphope Burn, in Channelkirk, is the true origin of
the Leader whose total length is a little over 21 miles.
Leaderdale and Lauderdale are but variants of the name.

The upper course of the Leader among the Lammermoors is through bleak monotonous hill scenery, but its middle and lower reaches pass into a fine series of landscapes—the "Leader Haughs" of old song. Lauder and Earlston are the only towns in the valley. Quitting Earlston, by the Rhymer's Tower, the river widens its course, flowing with fuller stream, gathering beauty as it goes. At Cowden-knowes and by Redpath's wooded braes, it is supremely beautiful. After a total descent of 1160 feet it falls into the Tweed at Drygrange.

The Whitadder, 24 miles long, seven of these in East Lothian, enters Berwickshire a little above Cranshaws. Running with rapid circuitous course through Cranshaws, Longformacus, and Abbey St Bathans, it turns southward towards Preston, then east, and joins the Tweed about 3 miles above Berwick. From source to mouth it falls over 1000 feet, and meanders through some of the most charming scenery of the Merse. The Blackadder rises at the base of the Dirringtons (1309 feet), flows south-east through Greenlaw, Fogo and Edrom, and joins the Whit-adder at Allanton, after a score of miles of numerous alternations of calm pool and brawling current. The Dye rises near Lammer Law, runs east through Longformacus, where the scenery presents many impressive river-effects, and after a descent of 1000 feet, unites with the Whitadder at Ellemford. A little above Longformacus it receives the tiny tributary of the Watch. The Eye, 20 miles in length, rises in Monynut Edge near Cockburnspath, runs in a south-easterly direction to Ayton, and after being joined by the Ale pursues a north-easterly course and falls into the sea at Eyemouth. The Leet, 12 miles long, is of Whit-

some origin. It flows through Swinton and Coldstream
where it beautifies the Hirsel and Lees policies before it
joins the Tweed.

Teviot and Jed are main arteries running through Roxburghshire. Beginning about Teviot Stone on the borders
of Dumfriesshire, the Teviot has its course wholly within
its native county. It is 40 miles in length, and when it
reaches Kelso has fallen 1200 feet. Ramsaycleuch Burn
and the Frostylee (formed by the junction of the Linhope
and Limeycleuch Burns) are its upland tributaries. At
Teviothead the river starts on its own, to run through a
land haunted with the romance of centuries, and to become
next to the Tweed the most besung of Border streams.
Its basin is at first a barren vale flanked by lofty green
hills. Then it becomes a strip of alluvial plain screened by
terraced banks clad with the richest vegetation, and with
long stretches of undulating dale-land, and overhung at
from three to eight miles by terminating heights. In its
lower reaches it is a richly variegated champaign country
possessing all the luxuriance without any of the tameness
of a fertile plain; and stretching away in resulting loveliness
to the picturesque Eildons on the one hand and the dome-
like Cheviots on the other.

At Newmill, the Allan Water, having run six miles from
its Skelfhill fount, is one of the first upland tributaries of
the Teviot. Further on, the river passes the massive white
pile of Branxholme, ancient key of the pass to the south.
Two miles on is the russet-grey peel of Goldielands near
the influx of the Borthwick with the Teviot.

At Hawick, Teviot and Slitrig meet. From Hawick to
Kelso the distance is 21 miles. Five miles below Hawick,

under the shadow of "dark Ruberslaw," lies the tidy little
village of Denholm, birth-place of two of Scotland's lin-
guists—John Leyden and Sir James A. H. Murray. At
Spittal the Rule joins up. A little below Chesters the
brown waters of the Ale pour into the Teviot and add
greatly to its dimensions. The river now winds through

Ruberslaw
In the foreground is a monument to John Leyden on the site
of Henlawshiel Cottage where his boyhood was spent

the domain of the Marquess of Lothian and passes in front
of Monteviot House. In about a mile it receives the Jed.
Two miles further on the Oxnam falls in. Teviot then proceeds along the valley till it arrives at Kirkbank, when it turns
north. Between its junction with the Kale at Ormiston,
and Roxburgh, it assumes a more rapid course, until passing
Sunlaws and the base of Roxburgh Castle, it "rolls upon
the Tweed."

Borthwick Water (16 miles) is formed by the union of Craikhope, Howpasley and Brownshope Burns at 1500 feet above sea level. Occasionally it forms the division between Roxburghshire and Selkirkshire. Its course is chiefly through Roberton, passing Hoscote, Deanburnhaugh, Borthwickbrae, and "wood-girt Harden" sentinelling a deep and narrow vale. Borthwick unites with the Teviot about two miles above Hawick. The Ale (24 miles) has its source in Henwoodie (1189 feet) in Roberton. After running for five miles, it flows through Alemuir Loch. It next drains the parishes of Ashkirk and Lilliesleaf, part of Bowden and Ancrum, and turning east and south, enters the Teviot about a mile from Ancrum village. Places of interest on its banks are "Riddell's fair domain," Linthill, Cavers-Carre, Longnewton, Belses, Kirklands, and Ancrum.

The Slitrig (13 miles) is formed by the junction of several streams above Shankend, one of them being the Lang Burn, issuing from a lochlet on the skirts of Windburgh. Running with quick descent northward through Cavers, Stobs, Kirkton and Hawick, the Slitrig joins the Teviot near the centre of the town. A wild, unruly river, it has been subject to considerable floods. One of these, in 1767, swept away fifteen houses in Hawick, with a corn-mill, "and the very rock on which they were founded was washed so clean that not a bit of rubbish or vestige of a building was left." Another, in 1846, left a similar tale of desolation.

The sources of the "mining Rule" (12 miles) are the Wauchope, Harwood and Catlee Burns, which join above Hobkirk. Sweeping round Hobkirk, it passes to the west

of Bonchester Hill and to the east of Ruberslaw, joining
the Teviot a little below Spittal. Its channel is mostly a
narrow glen cut through the freestone rock and overhung
on both sides with woods. Wauchope and Harwood nestle
beside their own burns. Hyndlee and Wolfelee are on the
Catlee Burn. Below Hobkirk are Bonchester Bridge,
Greenriver, Hallrule, Wells, and Bedrule. Rulewater is
associated with the turbulent Turnbull clan, one of whom
is said to have saved the life of Robert the Bruce from an
attack by a wild bull or bison. Hence the surname of his
tribe.

The Jed rises between Needlaw and Carlintooth on the
Liddel border. It intersects or bounds the parishes of
Southdean, Edgerston, Oxnam, and Jedburgh. Its general
course is east and north and its length to Jedfoot about
17 miles.

The Oxnam (10 miles) rises between Jed and Kale, and
running north through Oxnam and Jedburgh joins the
Tweed below Crailing. Plenderleith, Riccalton, Oxnam
Kirk, Crailing Hall, and Crailing House are places of note.
The Kale rises in Fairwood Fell (1230 feet) close to the
Border line, on the northern slope of the same range which
sends the Tyne, Breamish, Coquet, and other rivers into
England. Running north-east and north through Oxnam,
Hownam, and Morebattle, it turns west, touches Linton,
passes Morebattle village, bisects Eckford, and reaches the
Teviot at Kalemouth, having made a descent of 1135 feet.
Places of interest on its banks are Towford, Smailcleuch,
the Chesters, Clifton Park, Marlefield, Grahamslaw,
Haughhead, and Mosstower. Cessford Burn is a rivulet
which joins the Kale between Morebattle and Marlefield.

On its east bank are the ruins of Cessford Castle, a strong-hold of the Kers. The Bowmont (12 miles, anciently Bolbent) is the resultant of many head streams whose source is the Cheviots, at an altitude of from 1500 to 2000 feet. Running north through Morebattle and Yetholm, past Belford, the ruined kirk of Molle, Attonburn, the two Yetholms, and Venchen, it crosses into Northumberland and joins the Till at Flodden Field.

The Eden starts as a Berwickshire stream in Legerwood parish. Winding through Westruther, Gordon, Hume, Earlston, Nenthorn, it enters Roxburghshire near the Pinchburn at Smailholm. Passing through Stichill, where it tumbles over a linn 40 feet high, it makes for Kelso and Ednam and glides into the Tweed at Edenmouth, after a total descent of 760 feet in a run of nearly 24 miles.

The Liddel rises near the sources of the Jed at an altitude of 650 feet. After a course of 27 miles with a fall of 540 feet, it joins the Esk at the Moat of Liddel below Canonbie, 12 miles north of Carlisle. It is fed by a score of affluents, the chief of which are the Hermitage and Kershope Waters, the latter constituting for about nine miles the immediate boundary between Scotland and England. Other tributaries are Peel, Dawson, Lariston, Black, Tweeden and Tinnis Burns. Hermitage Water has its origin in a combination of Billhope, Twiselhope, Gorranberry and Braidlee Burns. Running east and south, it is joined by Hartsgarth Burn, and after a course of 12 miles enters the Liddel at Westburnflat.

The Lochs in our counties are few and small. Coldingham Loch covers 30 acres. Though only 300 yards from St Abbs Head, it is 300 feet above the sea, and is of considerable depth. Beautiful lochlets exist within the policies of Duns Castle—the "Hen Poo"—Spottiswood, and Mellerstain. Yetholm Loch is about a mile and a half in circumference. Primside is being steadily drained off. Hoselaw Loch covers 33 acres, and was probably connected with its neighbour, Linton Loch, now drained. Cauldshiels Loch, bounding Abbotsford estate, is a beautiful sheet of water, in area not more than a mile. Out of it flows Huntly Burn down to the Rhymer's Glen —a misnomer, Scott supposing this to be the scene of True Thomas's tryst with the Fairy Queen.

6. Geology and Soil.

To understand the essential features of the soil, the scenery, and agriculture of a county we must know something of its geology. By geology we mean the study of rocks. The term *rock* is applied to the soft as well as the hard material of the earth's crust, to sand and clay as well as granite and sandstone.

Generally, rocks may be classified into two divisions: (1) sedimentary; (2) igneous. A third division, known as metamorphic rocks, does not occur in our district.

Water-formed, or sedimentary, or aqueous rocks have been laid down under water in beds and layers—millions of years before the appearance of man on our planet. This process of formation is seen going on at the present day in a lake or at the sea-shore where mud or sand accumulates in layers, and in time by the superabundant weight becomes converted into rock. Originally all beds have been almost horizontal, but in course of time owing to earth movements they may have become tilted up and may now lie at any angle. The amount of slope from the horizontal of such rocks is called its *dip*. Portions of inclined beds of rock appear on the surface. These are termed the *outcrop* of the beds. On account of internal movements within the earth its crust is slowly undergoing contraction and sometimes rocks yield to pressure by *faulting*. Igneous or volcanic rocks have been poured out on the earth's surface in extensive sheets when in a molten condition; or the molten matter may have been injected between beds of rock, or thrust up through beds, cutting them for many miles, as in the case of the Yettan Dyke from the coast of Northumberland to the Border, or again it may have cooled deep down in the earth to form what we now call granite.

That volcanoes existed in the Southern Uplands of Scotland from early geological time is quite well proved. Many of their *necks*, that is, plugged-up rents, are distinctive landmarks in the counties. The rock of these necks is hard and withstands weathering much better than the surrounding adjacent rock.

There are five main divisions of the Geological Record:

1. Archaean, containing few traces of organic life.
2. Palaeozoic or Primary, containing Ancient Life Era.
3. Mesozoic or Secondary, containing Middle Life Era.
4. Cainozoic, or Tertiary, containing Recent Life Era.
5. Quaternary or Post-Tertiary, containing Man.

These divisions are based on the fossil remains found in the rocks.

The earliest visible record of the geological history of

Berwickshire and Roxburghshire begins with rocks of the Ordovician and Silurian Age. These are followed, unconformably, by the Lower Old Red Sandstone, and these again, also unconformably, by the Upper Old Red Sandstone, the record being closed by strata of Lower Carboniferous Age, following conformably those of the Upper Old Red Sandstone. The exposures of all these formations are much obscured by the glacial deposits, which cover a large surface of the land. There can be no doubt that geological formations of a much later date once covered all that is now visible, but they have been removed by subsequent denudation, and what we now see is only the remnant of a once much more complete geological sequence. The accumulated depth of the deposits amounts to many thousands of feet and the space of time represented is incalculable.

At the present time our earliest rocks are in an extremely contorted and folded state, and where they do not appear at the surface underlie all the newer formations. They extend across Scotland from east to west and comprise the whole of the Southern Uplands. They are marine in origin. The lowest beds are of Arenig Age, and have at their base an extensive lava flow (Andesitic), called from its peculiar formation, a pillow lava. This is succeeded by beds containing deposits of Radiolarian chert, which have been shown to have a special relationship to the pillow lavas. There are many bands of Graptolite shales containing characteristic species of graptolites, distributed in such a way that individual beds can be traced by their means through all their foldings and inversions. In the higher beds of the series trilobites are met with. Much of the strata is very barren of fossils, which are found sufficiently abundantly in special localities only. These Silurian rocks, largely composed of hardened mudstones with bands of shale, limestone, and grits were deposited at the bottom of a sea which covered the whole area. When deposition ceased, certain movements took place in the earth's crust, and enormous lateral pressure accompanied by upheaval resulted in intense folding of the strata. As soon as the land appeared above the sea, denudation began and carried away a very large amount of the deposits before subsidence again brought the land down to, or below, sea level, when another series of deposits was laid across their upturned edges.

Silurian rocks cover about a quarter of the area of Berwickshire and Roxburghshire, and are to be found in five areas. The first and the principal area lies along the western borders. The second, separated from the former by a narrow band of Upper Old Red Sandstone, forms the southern part of the Lammermoor Hills. The third lies to the east of this, and its coastal margin forms the magnificent cliffs extending south-eastwards from Siccar Point to St Abbs Head. This section, particularly at Siccar Point, is much visited by geologists. Below the cliffs can be seen one of the best exposures of a violent unconformity. The nearly horizontal layers of the conglomerates and sandstones of the Upper Old Red Sandstone are observed lying across the vertical edges of the Silurian Rocks. Another very good exposure of this unconformity occurs at Cockburn Law north of Duns. Further along the coast, by Fast Castle to St Abbs Head, the great folds in the Silurian strata are very evident in the face of the cliffs.

The fourth area forms a triangular space which extends southward from Eyemouth and Burnmouth to the suburbs of Berwick. The fifth and smallest area rises to the surface from beneath the Cheviot lavas on the south-west flank of the Cheviot Hills, and about two or three miles to west and south-west of Jedburgh. In only one small area are Graptolitic shales found—on Soutra ridge near Channel-kirk.

The Silurian rocks are succeeded by the Lower Old Red Sandstone. The lowest non-volcanic deposits are now very poorly represented in Berwickshire and Roxburgh-shire. At one time they must have covered a large area, but were subsequently removed by denudation. The only important area remaining is in the neighbourhood of Reston, Coldingham, St Abbs Head, and Eyemouth, where they occur as outliers. The volcanic series of the Lower Old Red is found in a large area extending eastwards from a line drawn from a point two and a half miles south of Sprouston, in the north, to Edgerston, in the south, right across the Border, where it includes the intrusive mass of granite forming the Cheviot Hills. This great series of lavas are of Porphyritic Andesite type (i.e. crystals of Andesite, a lime soda felspar, are scattered through their mass). They are very vesicular in many parts, the vesicles, or old steam holes, now being filled with infiltrated mineral matter. Similar intrusions are found at Cockburn Law, and Priestlaw in Lammermoor.

Succeeding the Lower Old Red Sandstone, the Upper Old Red is largely represented in both counties. It stretches in a continuous strip from Dunbar southward past Duns, Greenlaw, Earlston, Melrose and Jedburgh to some distance south of the western margin of the Cheviots. A branch of this strip runs eastwards past Chirnside, nearly to Berwick, where it is cut off by a great fault. Volcanic activity appears to have ceased during the whole time of the deposition of the Upper Old Red. Fossils are very rare, and only fragmentary, such as scales of the fish *Holoptychius nobilissimus*, *Bothriolepis obesa* and *Palæopteris hibernica*. Specimens of the first of these from the Black Hill Quarry at Earlston (discovered in 1866 by members of the Berwickshire Naturalists' Club) and of others from Melrose, Siccar Point, and Rule Water, may be seen in the Geological Gallery of the Royal Scottish Museum.

Soon after the close of the Upper Old Red period there followed a great outburst of volcanic activity, the effects of which are well seen in a series of rocks of basalt lava (basalt is a dark-coloured igneous rock which when molten is very fluid) called the Kelso Traps. These can be traced southwards from Greenlaw across the Tweed west of Kelso to the northern flank of the Cheviots. They agree closely in age with the volcanic rocks of Arthur's Seat, Edinburgh, but are probably somewhat earlier. Besides these basalt lavas, there are a great number of volcanic necks, and intrusions of basalt as well as of different varieties of Trachyte (a lighter coloured igneous rock which is viscous when molten). A notable example of the latter is the much degraded intrusive mass of the Eildons.

Great interest centres round the way in which the extensive denudation of a district so permeated with hard igneous rocks has brought about the present features of the scenery. Sir Archibald Geikie, in his *Scenery of Scotland* says: "The best locality for tracing the influence of such igneous rocks upon landscape is in the tract of Border country between Birrenswark and the Merse of Berwickshire. Among the more prominent eminences of this kind are Tinnis Hill and Watch Hill in Liddesdale, Pike Fell and part of Arkleton Fell in Ewesdale, Greatmoor, Maiden Paps, Scarod Law, Leap Hill and Windburgh Fell at the head of Slitrig Valley, Bonchester Hill, Ruberslaw, Black Law, Dunian Hill, Lanton Hill, Minto Crags, Peniel Hill, the Eildon Hills, and the numerous little hills between Maxton on the Tweed, and the foot of Lauderdale, and the long line of craggy heights that stretch by Smailholm and Stichill to Greenlaw."

We come now to the highest geological series represented in our shires—the lower members of the Carboniferous. The lowest deposits, called the Ballagan Beds, occupy a large surface in the two counties, in two areas: one in Liddesdale, the other in the Merse. The sedimentation was probably partly in lagoons, in estuaries, and even in open sea, as would be inferred from the different kinds of deposits encountered. Sun-cracks are common, even records of rain-prints are preserved, and deposits of gypsum are found. Very fine plant remains have been collected in the Coldstream district.

In the Liddesdale area, towards the south-west, west, and north-west, bands of Fossiliferous Limestone occur; they abound in marine brachiopods, corals and other fossils indicating marine conditions of sediment.

The strata succeeding the Ballagan Beds are the Fell Sandstones, composed mostly of sand with occasional beds of gravel. They occur almost entirely over the Border, and are succeeded by the Scremerston Coal series. Of this latter series a small strip is found on the shore, reaching from Berwick to Burnmouth, owing their occurrence in that place to a powerful fault which brings them down against the Silurian rocks to the west.

A geological feature belonging to Tertiary times is the intrusive dyke of dolerite which runs across Roxburghshire, nearly in a straight line from Rothbury in Northumberland, through Hawick to Hellmuir in Selkirkshire.

The whole surface of the two counties bears unmistakable evidence of the work of land ice during the great ice age. Let any one examine the one-inch map of the Ordnance Survey, where the trend of hill and valley is well defined by hill shading. It at once becomes apparent from the parallel nature of the grooves, that the great ice sheet moved in certain definite directions. The ice coming from the Highlands, and moving eastward along the valley of the Forth was met by the ice flowing down from the range of the Cheviots, so that over a large part of the two counties the direction of flow was roughly from south-west to north-east. The two streams coalesced just south of St Abbs Head, where the course of both was altered by meeting the flow of ice from Scandinavia, across the North Sea; when all three streams, carrying their load of boulders and other material turned to the south-east or south-south-east along the coast of Northumberland. The Boulder Clay, variously coloured by the characteristic rocks of each district, is found everywhere; as are also those curious long mounds of sand and gravel called Kaimes, which are due to the action of the melted waters of the retreating ice sheets. Border tradition says that the Kaimes are the different strands of a rope, which a troublesome elf was commanded to weave out of sand. The strands were all prepared, but when the imp tried to entwine them, each gave way, hence the broken parts of the Kaimes have remained to this day.

Both our counties show great diversity of soil. Clay, loam, sand, peat, and friable earth are the chief constituents. As a rule each of these distributes itself over extensive tracts of land and is less broken than in many counties. Only occasionally do we find every variety of soil attached to a single farm. Clay is the common ingredient of the "Howe o' the Merse"—the central and richest part of Berwickshire. Black in colour and of good depth, it is extremely fertile and is much more productive than the Roxburghshire clays. There are approximately 45,000 acres of clay lands in this county. A fine loamy soil is the characteristic of Leaderside, the banks of the Whitadder and Blackadder, of lower Teviotdale, and all along the Tweed from near Dryburgh. It may be said to constitute the whole area between Tweed and Teviot, and is therefore the most fertile soil in Roxburghshire. Peat, with the exception of the mosses at Fans, Gordon, Dogden, and a few lesser bogs, is seldom found in the lower lands of either shire. It is chiefly confined to the uplands, to the outskirts of the counties. In Liddesdale it is singularly interstratified with the clay deposits. Deep peat moss is not common. Sand is not general. Upper Teviotdale shows sand and gravel. In Westruther and some other parts there is black sand having the appearance of peat. What may be termed friable mould covers a large area very suitable for barley production. Upper Lauderdale and Rule Water are chiefly of this character.

7. Natural History.

In early times the island of Britain formed a part of the Continent of Europe. There was no North Sea but a vast bridge of land united all those countries together. Across this flowed the great rivers, the Rhine, Elbe, Ouse, Thames, Tay and Tweed. In course of ages the land underwent a gradual subsidence, the bridge disappeared, and Britain became surrounded with a shallow sea, at the most not more than 600 feet deep. Before this there had crossed from Central Europe the progenitors of those animals and plants whose descendants are still the flourishing fauna and flora within our shores. All continental forms of life did not find their way across, however, for Europe has a far greater variety than Britain, while in Ireland there are even fewer, the theory being that Ireland was cut off from the mainland at a much earlier period. This earlier separation no doubt explains the limited mammal-life of that country, as well as the proverbial paucity of its reptilian life. Britain itself has only about forty species of land animals contrasted with the ninety common to the continent, which of course include our forty. Curiously enough, our red grouse, denizen of every moor in Scotland, is the only bird unknown to the European continent.

For the student of Natural History the Border counties offer a field of deep and undiminished delight. The Berwickshire coast-line, the Tweed and its tributaries, the numerous woods and moorlands, the dells and dingles of the hills, have each their characteristic forms of plant and animal life. It is impossible to do more than indicate a few of the leading features.

The Fauna of Berwickshire and Roxburghshire is that which obtains throughout Scotland generally. In the class of Mammalia, or suckling animals, three out of six species of bats are known to occur in our area—the long-eared bat, the pipistrelle or common bat, and D'Aubenton's bat. The latter, however, is only occasionally met with. Among the Insectivora, or insect-eating animals, the hedgehog is fairly plentiful, chiefly in the neighbourhood of cultivation. The mole is ubiquitous. The common shrew is abundant. The lesser shrew—a rarer type—was not recorded in the counties until 1889. The water shrew—a difficult animal to observe—has been seen on the Berwickshire coast, and at Abbotsford. Of Carnivora, or flesh-eating animals, the wild-cat (*Felis sylvestris*) is extinct; probably the last specimen being killed at Wolfelee in 1880. The wolf, common of old in the forest region and hills, disappeared from Teviotdale and the Cheviots about the middle of the 18th century. The polecat, colloquially the "foumart" from its intolerable stench, ceased to exist in the Border district about 1850. The marten, too, has passed from its ancient haunts, but the stoat or ermine and its smaller congener, the weasel, are still common. Berwickshire and Roxburghshire are noted in the annals of fox-hunting. Foxes (*Vulpes vulpes*) are plentiful and show little sign of decrease. The otter is being gradually hunted to its extinction, though it is fairly common on the Whitadder and Teviot, and in the Cheviot district. The badger, an animal formerly abundant, as shown by many place-names beginning with *Brock* (e.g. Brockholes), is still far from rare. Of the order Rodentia, or gnawing animals, the squirrel has a wide distribution all over the counties. Only one specimen of the harvest-mouse has been recorded, but the long-tailed field-mouse is plentiful, whilst the house-

Blue or Mountain Hare
$\frac{1}{9}$ *natural size*

mouse and the brown rat are pests everywhere. The old British black rat has long been exterminated. The water-vole, the short-tailed field-vole, and the bank-vole are abundant. The rabbit is ubiquitous. The common hare, since the passing of the Ground Game Act of 1880, has become less general. The blue or mountain-hare is frequently seen.

Centuries ago, when our counties were in their wildest state and extensively afforested, red deer (*Cervus elaphus*) were abundant in many localities, but as truly wild animals they have been extinct for more than 200 years. Their remains have been exhumed from many peat deposits, and in digging operations—at Earlston gasworks, Coldingham, Linton, Maxton, Whitrig Bog, and Westruther. Herds of fallow-deer exist at Carolside, near Earlston, and at Ancrum, and roe-deer have been reported from the Pease Dean in Berwickshire. Bones of the wild ox (*Bos primigenius*) have been found at Whitrig Bog, and at Swinton in Berwickshire, at Jedburgh, Lilliesleaf, and Linton Bog in Roxburghshire. Bones of the beaver (*Castor europæus*) were found in Middlestots Bog, in October 1818, and an almost perfect skeleton was dug up in Linton Bog at a later period. In a sea-bordered county like Berwickshire seals are occasionally observed from the coast, especially the great grey seal. The porpoise is not uncommonly seen by fishermen along the whole length of the coast. Two examples of the white-beaked dolphin are recorded to have been caught near Berwick-upon-Tweed, one of which was secured for the museum at Kelso.

Both counties present the most excellent opportunities for a study of bird life. Hill and valley and the banks of every stream bear witness to the variety and abundance and charm with which Nature thus makes her appeal. The total number of distinctively British birds has been estimated at 410. Of these at least 300 species have been reported from the Border at one time or another. The author of the *Birds of Berwickshire* has enumerated 188 varieties occurring in that county, without reckoning those whose habitat is the coast line. The *Aves* of Roxburghshire are similar to those of the sister shire.

It is impossible to attempt a complete list here, and only typical examples can be given. Everywhere amid the blush of Spring is heard the song of the thrush or mavis—"the throstle with his note so true." The missel-thrush, the red-wing, and the field-fare are now comparatively common. The ring-ousel or hill blackbird is a characteristic up-lander, appearing early in April and leaving again in autumn. The stone-chats, redstarts, and wheatears come with the early summer, the latter locating themselves mostly on the moorlands, though found also on the coast at St Abbs Head, and between Coldingham and Eyemouth. The redbreast is found in all suitable places. The willow-wren, wood-wren, golden-crested wren and the various warblers are common. The dipper disports itself by most of our rivers. The great titmouse, the coal titmouse, and the blue tit are abundant. The pied wagtail is plentiful, but the white wagtail has only been observed on two occa-sions—at Nenthorn and Eyemouth, and the yellow wagtail, always rare in the Border country and in Scotland, has only been seen once. The rock-pipit is not uncommon on the rocky shore-line between St Abbs Head and Berwick. Records of the great grey shrike are fairly numerous all over our area, and instances of the red-backed shrike—a rare visitant—have been reported from Cavershaw, near Hawick, and Byrewalls, near Gordon. The spotted fly-catcher and the pied flycatcher are fairly frequent. The greenfinch, goldfinch, chaffinch (shilfa) and bullfinch are all found. The siskin is perhaps more common in Berwickshire than in any of the other Border counties. The linnet, the lesser redpole, and the crossbill breed freely throughout the district, and the rare mealy redpole has been reported at Harelaw, in Berwickshire. The corn-bunting, the yellowhammer, the black-headed bunting, the snow-bunting are all fairly plentiful. The starling was almost unknown in the counties about the beginning of last century: now it is seen everywhere. The chough (*Pyrrhocorax graculus*), or red-legged crow, an uncommon bird, is said to have bred at St Abbs in 1895. The jay, after an almost total disappearance, is on the increase in our counties. The magpie is only occasionally seen. The jackdaw breeds abundantly on the cliffs at the coast and on inland crags. The raven now confines its operations to the wildest parts of the hills, where a few pairs lead a precarious existence; a pair nested in Lauderdale in 1921. Carrion and hooded crows have their strongholds among the Cheviots. In 1887, the total number of rookeries in Berwickshire was 66, and there were 88 in Roxburghshire. The lark is a resident in all parts of the Tweed area. Of the owl tribe, the barn-owl, once plentiful, is now a mere straggler in the counties. The long-eared owl is still fairly common, but the short-eared owl is only occasionally found. The brown or wood owl is resident and widely dispersed throughout our area. Of Falconidae, only one specimen of the marsh-harrier has been reported—at Ayton in 1903. The hen-harrier is a straggler. The common buzzard and the honey-buzzard are occasionally seen. The golden eagle has ceased to breed in the Border district. The sparrowhawk is rapidly decreasing in numbers. The kite has quite disappeared. The peregrine falcon, or hunting hawk, still breeds at Fast Castle, and a nest has been reported at Ruberslaw. The merlin holds its own in the higher hill districts. The kestrel is fairly common on the Berwickshire coast. Specimens of the osprey have been shot in both counties, but it is now comparatively seldom met with. The common cormorant is found throughout the whole of the coast area. The green cormorant is less common. The gannet or solan goose, locally known as the "Bass Goose" has been found as far inland as Chapel-on-Leader and Spottiswood. In 1882 there were in Berwickshire eight heronries and four in Roxburghshire. Favourite haunts of the bittern, or "bull o' the bog," were Billie Mire and Huntlywood Moss. Probably the last specimen to be recorded was shot at the mouth of the Whitadder in 1890. Of the Anseres or goose tribe, the bean and pinkfooted goose are common, and may be seen crossing the country in large and sometimes immense flocks. The barnacle, brent, and Canada goose have been occasionally encountered. The wild swan or whooper, and Bewick's swan have been brought down by local gunners. The mallard or wild duck is common. The shoveler has been observed frequently. The teal breeds on many lochs and moors of the Border. The wigeon, the tufted duck, the scaup duck, the golden-eye, the eider duck, scoters, goosander, and smew have all been noted. The ring-dove or wood-pigeon, stock-dove, rock-dove, and turtle dove are recorded. The common moor birds—golden and grey plover, lapwing or peewit, blackcock and grey hen, red grouse, curlew or whaup, are abundant. The common snipe is plentiful, the great snipe rare. The common sandpiper is found by every river. Two records only occur of the crane—in 1803 near Hawick, and another at Threepwood, near Lauder. The dotterel is not uncommon. The pheasant is found in all suitable woods. The partridge is plentiful. The quail is rare. The landrail or corncrake, the water rail, the water hen, and the coot are common. The great bustard, mentioned by Boece in 1526, has long since disappeared from the Border. The great skua is rarely met with, though its congener, the arctic skua is found along the whole coast-line. Other birds of the coast are the razorbill, the common gull, the black-headed gull, the common guillemot, the kittiwake, the puffin, the redshank, the great northern diver, the little auk, the great crested grebe, and the horned grebe. The little grebe breeds in both counties. The red-necked grebe and the black-necked or eared grebe are rare. The storm-petrel is an irregular straggler to our area.

The number of British reptiles does not exceed six. Three belong to Scotland and all are found in our counties. The common lizard, though rare to the Border, has been met with at Cockburnspath, Quixwood and Gordon, and amongst the Roxburghshire Cheviots. The slow-worm or blind-worm occurs at Grantshouse, and in the west of both counties it is found in small numbers. It is not a snake in the ordinary sense but timid and perfectly harmless. Adders are plentiful and within recent years large specimens have been killed at Gordon Moss and amongst the Lammermoors. The common frog and toad may be looked for in all suitable places. The warty newt, and the smooth newt or "ask," occur in several places. The palmated newt is unknown.

All the common Lepidoptera are found in the counties, and many rare varieties have been recorded in larvae and on the wing. In 1895, 360 varieties were noted in the Hawick district, including the uncommon *Scopula olivalis*, found in Denholm Dean, and *Lomaspilis marginata*, found at Ruberslaw and Humbleknowes. Three hundred varieties have been noted from Berwickshire, including *Mamestra albicolon*, the white colon, the rarest capture in the county. Another very rare butterfly is *Vanessa antiopa*, or Camberwell Beauty. Dragon-flies are plentiful at Mellerstain and the Death's Head Moth (*Acherontia atropos*) has been got at Earlston, Lauder, and other places.

A collection of Spiders (Araneidea) and Harvest Men (Phalangidea) made at Eyemouth in 1895 numbered 74, the most noteworthy being *Dysdera crocata*, which had not previously been reported in Scotland. The Coleoptera of the counties has been described in various papers of the Berwickshire Naturalists' and other clubs. Seventy specimens of land and fresh water Mollusca have been recorded.

In their floral life our counties may be regarded as one, they are so akin—the sea-board excepted. Many pleasant highways and byways intersect the shires. Everywhere, except in the uplands, we meet with smiling hedge-rows which always give grace to a countryside. These are mostly of hawthorn, but beech and privet are used. Mixed with these we find growing the wild raspberry, the bramble, and the sloe: honeysuckle, convolvulus, dog-rose and sweet-briar climb over them: on the banks are the wild straw-berry, wood sorrel ("cuckoo's meat"), forget-me-not, wild thyme, wood-vetch or wild pea, and the ubiquitous robin-run-the-hedge. In fields and meadows, in their proper season, are to be found abundance of wild violets, crowfoot, field lady's mantle, marguerite ("horse-gowan"), bluebells, speedwell, milfoil or yarrow (from which a medicinal tea is still made), buttercup, meadow-sweet and poppy. In the woods are the wood-anemone, wood crane's bill, the lesser celandine, wild hyacinth, primrose and foxglove: or the woods may be merely a sober green with carpet of male and lady fern and bushy bracken. A stiff roadside soil will yield varieties of campion, the centaury (*Centaurea Scabiosa*) or "horse's knots," better known on the Border as "soldier's buttons." In marshy soil flourish bog asphodel, bog-myrtle and cotton-grass (at one period largely cultivated, memories of its wide-spread usefulness remaining in the "lint-holes" of numerous parishes and in place-names like Lintlaw and Linthill), butterwort, ragged robin, marsh marigold, marsh violet and valerian. In ponds and ditches we find watercress, the bulrush, *Scirpus lacustris* (common at Mellerstain), iris, smooth naked horsetail, and spearwort. By burnsides and in deans we get garlic (abundant at Red-path Dean), coltsfoot or "dishylagie," and butterbur or "wild rhubarb," the largest leaf of any British plant. On walls, rocky places, and crumbling ruins we find the wall-flower, wall-rue, spleenwort and parsley ferns. Over many a Border cottage climb honeysuckle, clematis and virginia creeper. On the drier heights we meet with the wild pansy, the tormentilla (called by the Cheviot shepherds the "ewe-daisy"—where it grows the rot in sheep is seldom encountered), and higher up, common ling (*Calluna vulgaris*), brown in winter but changed to imperial purple when it breaks into flower in August. Borderers distinguish between bell-heather as "she" heather, and ling as "he" heather. Here, too, we encounter relics of a colder epoch—

blaeberry, cloudberry (*Rubus chamæmorus*), whortleberry
(*Vaccinium myrtillus*), thistle and juniper—alpinists all,
which come first of the four classes into which the flora
of Britain has been divided, the order being (1) Alpine,
(2) Sub-Alpine, (3) Lowland, (4) Maritime.

Every moist place will reveal its attendant beauty to the
eye which can see it—grass of parnassus, masses of sphag-
num moss, round-leaved sundew (*Drosera rotundifolia*),
saxifrages of various species including the rare *Saxifraga
hirculus*, which Dr Thomas Brown, a Berwickshire
minister, found near Langton Wood and had the good
fortune to add to the Flora of Scotland, and March stitch-
wort (*Stellaria palustris*), a unique treasure of Gordon
moss. Whinbloom is to be seen in all its golden fire on
many a Border brae, and the yellow broom has been en-
shrined in imperishable Border poesy.

Of Ferns (*Filices*) there are 24 varieties in the counties—
the polypody which is abundant, beech, oak, the bucklers,
the spleenworts, adder's tongue, hart's tongue, bladder,
prickly shield, moonwort, and rare on the Borders, the
royal fern (*Osmunda regalis*), seen at Spottiswood, where
Lady John Scott placed a strong iron cage over several
specimens to preserve them from the public. Of Willows
(*Salix*) there are 60 varieties, all more or less common.
Fifty-eight varieties of Grasses have been enumerated, 46
of these belonging to Lauderdale. Sedges number 41, including 34 of *Carex*. *Carex pallescens*, though found in
neighbouring shires, is absent from ours. *Equisetum* has
6 varieties, *Lycopodium* 4, Musci 154, Hepaticae (liver-
worts) 37, Lichens 97, Fungi 453, Marine Algae 105,
and Freshwater Algae 76.

Although Berwickshire and Roxburghshire are now so beautifully wooded, down to the middle of the seventeenth century the district was almost wholly destitute of trees. The great natural Forests of Lammermoor, Cheviot and Jed had vanished, and as afforestation was an unknown art the want was not speedily supplied. But the revival came and with it a transformation of the landscape into a thing of beauty. There are no forests in the ordinary sense, but wood is plentiful, especially in the low-lying districts and in the vicinity of the great houses, whose spreading parks are adorned with valuable timber of all types. Many shapely trees fringe the river-banks and the roadsides. The woods at Mellerstain, Carolside, Cowdenknowes, Marchmont, Thirlestane, Floors and Ancrum have long been famous. Some of the first larches in Scotland were planted at Minto. Duns Castle grounds contain picturesque specimens of oak, lime, sycamore, chestnut, ash, larch, elm, poplar, alder, beech, maple, willow, walnut, yew, elder, hawthorn, holly and hazel. Its avenue of 300 yards of lime-trees is one of the finest in the Border.

Many remarkable trees have been recorded. A gigantic wych-elm at Ednam measured 20 feet at 3 inches above the ground, and 18 feet 6 inches at 2 feet above the ground. It was one of the largest trees of the kind in the county. A lime at Edgerston measured 19 feet in girth and in a blown ash the number of rings counted was 300. A silver fir at Nisbet, in Berwickshire, is 100 feet in height. Another, at Carolside is 100. A third, at Mellerstain, is 109 feet. A very valuable oak there is 84 feet high with a clean bole of 47 feet, 8 feet in girth at 4 feet, and 11 feet at 5 feet. Several beeches also rise to a height of over 110 feet, and a magnificent larch, over 200 years old, is 93 feet. At Marchmont a group of Spanish chestnuts ranged in height from 66 to 102 feet. A silver fir has a height of 104 feet, and a beech has over 100 feet. Several fine specimens of the common yew of considerable size and of great age are within these grounds. The yew trees at Dryburgh are believed to be coeval with the abbey. The Covin Tree at Bemersyde was planted by a Haig seven or eight centuries ago. The Capon Tree (still showing ample, umbrageous features) and King of the Wood, both oaks, last survivors of "Jedworth's Forest wild and free," are the most ancient of all the trees of the Border, being probably 1000 years old. The first mentioned measures 26 feet 6 inches in girth above the roots. It is divided into three limbs, their girths being respectively 16 feet 7 inches, 11 feet 3 inches, and 10 feet 9 inches. It covers an area of about 90 feet. The second is 78 feet in height, 16 feet 6 inches in girth, and at 30 feet from the ground it is 11 feet 3 inches.

8. The Berwickshire Coast.

The Berwickshire coast-line, exclusive of minor indentations, is about 19 miles in length. Beginning at the eastern extremity of Mordington it proceeds in a north and north-westerly direction to St Abbs Head, which is about half the entire distance. Thence it trends almost due west by Fast Castle to the mouth of Dunglass Burn at the boundary of East Lothian.

Everywhere, with few exceptions, the coast is rocky and precipitous, perilous to foot-passengers, a menace to sea-farers. These bold, craggy bluffs range in height from 100 to 500 feet, the highest cliffs on the eastern sea-board of Britain. They possess extraordinary picturesqueness. Landing-places are sparse. Only Eyemouth can shelter and anchor a good-sized ship. The bays at Burnmouth, Coldingham and Cove are accessible for fishing-boats and very small craft.

Three miles out from Berwick-upon-Tweed, upon the Great North Road, stands what used to be Lamberton Toll at the confines of the shire where it marches with the Liberties of Berwick. Lamberton, dedicated to clandestine marriage-makings, was the Gretna Green of the eastern Border. More historic is the site of the Kirk of Lamberton. A deserted graveyard alone is left. Here, in 1503, the child-princess, Margaret Tudor of England, bride of James IV, was handed over to the care of the Scottish Commissioners. In its tiny rock-walled bay nestles the fishing hamlet of Ross. Invisible too from the highway red-roofed Burnmouth hugs a mass of cliffs close to the beach. Almost a replica of a Cornish village, it figures on many an artist's canvas. The highway itself is perhaps the most picturesque in the shire, with its majestic sea-scapes, wide waving coast-line as far as the Farnes and the Fife shore, and sweep of inland panorama. The whole of the Merse lies at one's feet. The distant Eildons are clearly seen. Here and there a narrow grass-grown valley slopes down to the iron-bound surf, ending in a precipice where the naked reddish-brown rock is visible. Broken Carboniferous and Silurian headlands stretch all the way to Gunsgreen. Each

The coast at Eyemouth

of these formations pierces the heart of the Border—the Carboniferous to the Solway, the Silurian to the Rhinns of Galloway.

Eyemouth, which sits snugly at the mouth of Eye Water, has been described as a town "dark and cunning of aspect, full of curious alleys blind and otherwise, and having no single house of any standing but what could unfold its tale of wonder." Lord Protector Somerset built a Fort here in 1547. It fell in 1560, but its name survives in a quaint rhyme:

> I stood upon Eyemouth Fort
> And guess ye what I saw—
> Fernyside and Flemington
> Newhouses and Cocklaw,
> The Fairy Folk o' Fosterland
> The witches o' Edincraw
> The rye rigs o' Reston
> And Duns dings a'.

Gunsgreen (in Ayton parish) overlooking Eyemouth harbour, had a sinister reputation in the smuggling epoch. A tragic event, never-to-be-forgotten in the annals of Eyemouth, occurred in 1881, when the major part of its fishing fleet succumbed to a terrific October gale, and 129 lives were lost. Bending round for a mile or two, we come to Coldingham Bay, the largest indent, where above an inviting sandy beach the caverned cliffs of St Abbs frown to a height of more than 300 feet. St Abbs Head is a magnificent mass of volcanic origin jutting out into the sea and joined to the mainland by comparatively low ground partly occupied by a large pond. Except at Cape Wrath or Dunnet Head there is no more weird, more awe-begetting scene, whether bathed in sunlight, or with white fogs floating about, or the storm raging in fury.

St Abbs

Innumerable caves have cut their way into the jagged cliffs, inaccessible by land and approachable by sea only at low water and in the calmest weather. All of them were

St Abbs Coast and Lighthouse

haunts of smugglers. Only the faintest traces are discoverable of the religious house reared by St Ebba upon this desolate headland in or about 670.

Modern St Abbs thrives by its trade in herrings and haddocks, and its successful catering for summer visitors. St Abbs Lighthouse, erected in 1862, flashes its guiding and warning light every ten seconds over 21 nautical miles.

After Dryburgh, Coldingham is the most sacred site in Berwickshire. Ecclesiastically, Coldinghamshire embraced one-eighth of the county. Edgar's Priory was a beacon light for religion long before Melrose and its compeers

Fast Castle

kindled their lamps. Petticowick Bay is a romantic little harbour at the base of the beetling cliffs. The names of Ravensheugh and Earnsheugh recall their departed inhabitants. Here the sea-wall rises sheer to 500 feet. Behind are Coldingham Loch and ling-covered Coldingham Moor. At Brander Cove the green cormorant has its home. Crossing the Dowlaw Burn—a botanical paradise—we reach Fast Castle mouldering on its lofty eyrie. The fastness of the iniquitous Logan of Restalrig, one of the Gowrie conspirators, it lives in literature with a better grace as the "Wolf's Crag" of the *Bride of Lammermoor*. Proceeding westward, passing innumerable caverns, stacks, heughs and carres, we arrive at Siccar Point. Close at hand and a little to the north of what is probably the site of the village of Aldcambus, and about three miles east of Cockburnspath, are the ruins of St Helen's Chapel. Of three sainted sisters it was said:

> St Abb, St Helen, and St Bey
> They all built kirks to be nearest the sea—
> St Abb's upon the Nabs, St Helen's on the lea,
> St Ann's upon Dunbar sands
> Stands nearest to the sea.

Why St Ann should appear in the rhyme instead of St Bey, is not clear. We now touch one of the most picturesque spots in the county—the Pease Burn with its romantic Dean. As a botanising centre the locality has a high reputation. Pease is literally "Paths." The diarist, William Patten, describing the Duke of Somerset's Expedition of 1548, says, "so steepe be these banks on eyther side, and depe to the bottom, that who goeth straight doune shall be in danger of tumbling, and the comer-up so sure of puffyng and payne, for remedie whereof the travailers that way have used to pass it not by going directly, but by paths and footways leading slopewise; from the number of which paths they call it (somewhat nicely indeed) the Peaths." Cromwell found it no less a barrier in 1650 when he spoke of it as a place "where ten men to hinder are better than forty to make their way." This great ravine lay in the route of the old Great North Road which ran by Press Inn and across the bleak heathy Coldingham Moor, almost hugging the coast for the last few miles of its Berwickshire journey. Pease Bridge was built in 1786. It is 300 feet long, 16 feet wide, and 127 feet high. The railway and the New Great North Road bridge cross the gorge further up. Cove Harbour, where there is a Coastguard Station, is a fascinating bit of coast scenery. At Dunglass Burn we move over into East Lothian.

9. Climate and Rainfall.

The word climate is held to signify the state of a country with respect to its weather. Weather may be defined as the character and behaviour of the atmosphere, and the climate of a country is accordingly the average of its weather conditions over a series of years. Its chief factors are temperature, pressure, and moisture of the air, the air's movement as wind, the rain or snow which it brings, the sunshine it permits—every modification of the atmosphere affecting man and animals and even the vegetable realm. These, again, are dependent upon (1) latitude, (2) position with regard to the great belts of high and low air-pressure, (3) distance, or proximity to the sea, (4) configuration of land surfaces, (5) altitude, and (6) to some extent, agriculture.

No Scottish county is large enough to have a distinctive climate of its own, though we sometimes hear it said that the climate of one district differs from that of another. That is not so. Each shares in the climate of the geographical region to which it belongs. Mere local differences arise through corresponding differences in the contour of the land, and its relation to the prevailing winds.

Scotland lies approximately between 55° and 60° North Latitude. As it faces the Atlantic with no land intervention to the Labrador coast—over 2000 miles—it is thus in the zone of prevailing westerly winds. Hence its climate is more uniform and oceanic than that which exists in England as a whole. It is the sea which really creates and influences the climate of Scotland. Latitude is a secondary consideration. Were it the main consideration we should experience a climate as rigorous as that of Labrador. Our temperature is principally due to the ameliorative action of the Atlantic, which is three degrees warmer than the air. During winter the Atlantic gives off its stored-up heat of summer. Driven by the winds from west-south-west, this operates benignly upon the great mass of the country, developing a much milder temperature upon the western portion than upon the eastern, where the conditions warrant us to expect a temperature more severe and extreme. As the surface of the North Sea is so much smaller, and its temperature only one degree warmer than the air, it will at once be apparent that even with its quota of stored-up heat, it cannot exercise the same benign influences. Nor must we forget that the warmer winds of the west are cooled and spent before the colder east is reached. At almost the centre of Scotland, for instance, we find a winter temperature of 38° compared with 40° on the west, and 39° on the east, while in summer the temperature of the east is generally a degree or two higher, say in July, on the Berwickshire coast than on the Argyllshire coast. For the same reason the east is drier than the west, the wind-borne moisture from that quarter seldom reaching the east coast. Continental influences are also a factor in the east.

Situated as Berwickshire and Roxburghshire chiefly are under the kindly shelter of the Lammermoors and the Cheviots, and the lesser heights on the west, both counties enjoy a climate of mild and temperate nature. The prevailing winds—evidenced by the growth of isolated trees—are from the west, but in spring cold easterly and north-easterly winds from the North Sea sweep across the slopes facing the coast-line, and are severely felt for several miles inland. On the other hand, the coast district benefits by the shorter duration of frost and snow. Although severe snow-storms may occur at any time from November to April, long-continued storms are of rare occurrence. The winters of 1879 and 1895–6 were the severest within living memory. Snow covered the landscape for months, and many of the rivers were ice-bound for a lengthened period. In the low country as one travels west snow is found to lie longer, and in the Lammermoors corn-sowing is not infrequently retarded. Nevertheless though snow may sometimes be seen on Cheviot upon Midsummer Day it required the ingenuity of Midside Maggie to produce a snowball from the Lammermoors on that date. Snow fell on 19 days in 1923.

The mean annual temperature of both counties is 46°: of summer between 56° and 58°: and of winter 37°. In both counties July is the warmest month. December and January are the coldest months in Berwickshire and Roxburghshire respectively.

The rainfall of Scotland is depicted on the map. It is expressed as the number of inches by which the land would be submerged if the rain of a year remained where it fell. The enormous difference between west and east will be

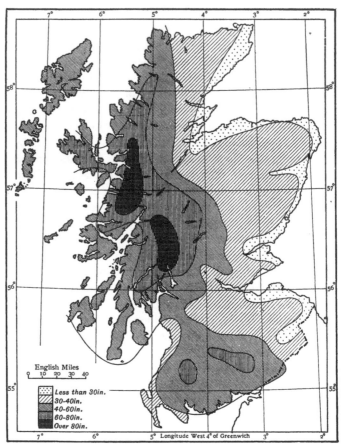

English Miles
0 10 20 30 40

Less than 30in.
30-40in.
40-60in.
60-80in.
Over 80in.

7°   6°   5° Longitude West 4° of Greenwich   3°   2°

Rainfall Map of Scotland
(after Dr H. R. Mill)

immediately apparent. The fact of real importance for us is that only some 9 per cent. of the total area of Scotland has a mean annual rainfall of slightly over 30 inches. This is a narrow fringe along the whole east coast from Caithness to Berwick. The west coast stations show such figures as 40, 45, 56 and 60, compared with 25, 27, 30 and 35 on the east, except in the region of the hills where the percentage is always higher. The rainfall of our counties, especially in their eastern parts, is thus amongst the lightest in the country. An examination of records kept at a number of stations shows the mean annual rainfall to be 32 inches and 30 inches in Berwickshire and Roxburghshire respectively, as against 47 for the whole country and about 32 for England.

The average for the last ten years at St Abbs Lighthouse (200 feet) was 24·05 inches: at West Foulden (250 feet), which may be taken as representative of the east end of the Merse, the rainfall was 25·78 inches. On Tweedside, Hirsel (94 feet) shows 26·60, and the centre of the Merse reports 26·29 at Swinton House (200 feet). Rowchester, at the west end of the Merse, has an amount similar to West Foulden. Duns Castle (500 feet) near the foot of the Lammermoors, shows a larger amount, 30·29, and at Rawburn (920 feet) in the Lammermoors, the figures averaged fully an inch more than at Duns Castle. In Lower Lauderdale, Cowdenknowes (360 feet) has 28·99. Burncastle (900 feet) in Upper Lauderdale has about four inches more, while Blythe Rig (1250 feet) on the hills above Burncastle shows an average of 38·55 for the last five years, being over five inches more than Burncastle rainfall for the same period. The rainfall in 1923 may be given for three of these stations: St Abbs Head Lighthouse had 25·57 for the year, the highest being November with 3·92, followed by August with 3·88, while the lowest was ·69 in March, followed by ·84 in June and 1·13 in July, August being an abnormally wet month. At Duns Castle the average for the year was 27·37, August being again the wettest month with 4·31 and June the driest with 1·00. At Cowdenknowes the yearly average was 26·75, with August 4·19, and June ·90. March was the driest month with ·87. September had 1·55, and January 1·74.

Returns from Roxburghshire for the period 1881–1915, show that at Wolfelee, in the Rule valley (604 feet), the average was 37·63: at Branxholme on Teviot (457 feet) 33·36: at Kelso (205 feet) 26·03: and at Jedburgh (229 feet) 26·89 inches.

The sun in Britain is above the horizon for about 4400 hours annually. In an average of five years Duns Castle records show 264 days of bright sunshine with 1190 hours; Marchmont had 1284 hours. For 1923 Marchmont reported 1238 hours and 289 days; Swinton 1083 hours and 275 days. May to September were the brightest months. December and January the dullest. On 96 days the thermometer fell below freezing-point.

Fogs are not frequent except for the occasional "haars" from the sea, which, as a rule, do not last long, and seldom reach far inland. Everywhere the air is pure and bracing. There is no contamination. The coast is cool and refreshing in the height of summer, and is much resorted to at St Abbs and Eyemouth. The like may be said of many of the upland districts in both counties, which are coming more and more into favour every year.

10. People—Race. Type. Language. Population. Occupations.

The counties of Berwick and Roxburgh, embracing the fertile valleys of the Merse and "pleasant Teviotdale," could not but prove attractive to any drifting or migrating people, and in point of fact the two shires seem to have provided a dwelling-place at one period or another for almost every race that settled in Britain. Thus we may assume that they were the home successively of representatives of the three races, the Mediterranean, the Alpine, and the Nordic. Of the primeval inhabitants of the district, as of Scotland generally, we know almost nothing; but we have traces of the Mediterranean people—the long-headed, dark, slimly-built men who came originally from North Africa, and reached our country by way of Spain—hence called the Iberian branch—and who have left deposits in the bones, weapons, etc., preserved in their long barrows, as also a slight but noticeable residuum on the physical character of the present population. Next came the Alpine race, by way of Central Europe—round-headed, dark, and of more substantial build than those whom they dispossessed. These again have left memorials of their presence in round barrows or tombs, but so far as can be recognised, appear to have had but little influence as an ethnological factor in Britain. Some investigators are inclined to call this Alpine people Celts—the true Celts, as distinguished from those who are now generally called Celts. For the latter, ethnologically, seem rather to belong to the third great immigrant race, the Nordic. When at length the obscurity of the primitive ages begins to give way before the dawn of history, we find the "Celts" (so-called) everywhere in Britain, and while they differed in language, customs, religion, from the Germanic tribes who came later, yet in physical characteristics the former were—and are— virtually identical with the latter. Both Celts and Germans were long-headed, prevailingly fair-haired and fair-skinned, tall and robust.

The Celt came in two waves—first the Goidel or Gaelic, then the Brython or Cymric man. Both can be traced in our counties, though about the beginning of our era the predominant element in the Scottish Border was a Cymric people, the Otadeni (or Otalini). Just about this period came the Roman settlement, but as the Romans did not intermarry with the native population, they left no permanent strain amongst our people. Then came a long series of immigrations from Nordic areas—Jutes, Angles, Saxons, and Danes, supplemented later by the Normans, who were originally pirates from Scandinavia.

Now, if we allow for migrations and intermixtures, the foregoing elements accord remarkably well with what we find to-day among the people of the Scottish Border The long head is common; tallness and vigour are noticeable features. The distinctively swarthy man is relatively rare; the ruddy much more frequent; but the great mass of the population present a mixture—an undecided brown, with a tendency to fairness, here flaxen, there auburn; in a word the Nordic with an infusion of the Iberian strain.

Traces of several of these immigrants are found in the place-names of the counties. Thus the outstanding physical features of the district—rivers and mountains—bear names of Celtic, mainly perhaps of Brythonic origin; as Tweed, Teviot, Eden, Ale, Elwyn or Allan, Eildon, Cheviot, Lammermoor, and several Pens—Penielheugh, Penchrise Pen, Skelfhill Pen.

Similarly, the older human settlements are Celtic in name: Kelso, Melrose, Earlston (i.e. Erchildun), Duns. The more recent village communities, again, generally bear English names: thus we have the numerous *hams*, sometimes disguised, and occasionally attached to an earlier Celtic name—Yetholm, Smailholm, Leitholm, Denholm, Coldingham, Ednam (Eden-ham), Edrom (Adder-ham), Midlem or Midholm: also the frequent *tons*, as for example, Paxton, Ayton (Eye-ton), Allanton, Sprouston, Oxton.

The Dano-Norwegian element is rarer, but we find many *fells*—Pike Fell, Carter Fell; a few *shiels* or *shields* as Cauldshiels; several *hopes*, as Sweethope, Hobkirk, Swinehope. The *klint* of Clinthill, Clintmains, is Norse for cliff.

The present-day vernacular of the Border Counties shows three fairly distinguishable types, those of (1) Berwickshire with North Roxburghshire, (2) Teviotdale, and (3) Liddesdale. The first corresponds generally to the Central Scottish of the Lothians, though it is characterised by a much broader vowel pronunciation. The Teviotdale type has been dealt with exhaustively by one of the greatest of recent philologists, Sir J. A. H. Murray, who, as a native of Denholm, knew the dialect root and branch. Dr Murray shows how this "Teviotdale" idiom covers also Selkirkshire and part of Dumfries, but stands apart from "Liddesdale," which is nearer the English dialects of Cumberland and Northumberland. Since Murray wrote, however,—say fifty years ago—curious changes have occurred. "Lothian" has crossed to Tweed from North Roxburghshire and effaced some of the earlier distinctive elements of "Teviotdale." That district is still the land of "yow" and "mey" (for *you* and *me*), and still preserves the *ee*-like termination in words like *peetee*, pity, *noteece*, notice (North Roxburghshire has *peetay*, *notace*). Other two pairs of diphthongs noted by Murray as found in such words as *clay-ee* (clay), *knō-oo* (knowe, i.e. knoll), and *nee-am* (name), *soo-ul* (sole), are not now heard in Teviotdale, but are current in Liddesdale and to some extent in the eastern part of Dumfriesshire. The last two forms are akin to the *niám, swōl* of Cumberland, but with change of stress. A grammatical point emphasised by Murray is the distinction between present particle and gerund, as *dancen*, *danceen* (dancing), and a corresponding distinction was made at an earlier day in North Roxburghshire, where the gerund was heard as *dansane*.

Mr George Watson, in his *Roxburghshire Word-Book*, has gathered together a vast and rich vocabulary of thousands of words belonging specially to Roxburghshire which show how well the dialect of that county has been preserved in its pristine purity and expressiveness, mostly, however, in the rural districts, where the influx of strangers has not been frequent and the influence of the town has not generally penetrated.

The Northumberland "burr" is not heard except around Berwick itself. Coldstream and Cornhill (immediately across the Tweed) are quite different in their manner of speech. At Cornhill the "burr" is everywhere.

The total population of Scotland on Census Day, 1921, was 4,882,288. Berwickshire had then 28,246 inhabitants, of whom 13,443 were males and 14,803 were females. In 1801 when the first census was taken the population of the county was 30,206, and from that date to 1861 each successive census showed an increase. The population in the latter year reached its maximum with 36,613, but since then each decennial census showed a decrease. In 1911 it was 29,643. Compared with 1861, the present population is 22·9 per cent. less, and compared with 1911, it is 4·7 per cent. less. There are 10 persons to each 100 acres. Of the persons enumerated 89·7 per cent. were returned as being of Scottish birth—17,243 were natives of the county. Sixty-five persons were returned as able to speak Gaelic as well as English, and none as speaking Gaelic without English. The civil parish of Coldingham had the largest population with 2830—only twelve more than Duns, the county town. Cranshaws had the smallest with 97.

Roxburghshire, with 44,989 inhabitants on Census Day, stands eighteenth in the list of Scottish counties. In 1801 the population was 33,721, and each successive census up to 1861 showed an increase. In that year the population reached its maximum with 54,119, but since then each census, with the exception of 1891, showed a decrease. In 1911 it was 47,192. Compared with 1861 the present population is 16·9 per cent. less, and compared with 1911 it is 4·7 per cent. less. Of male population Roxburghshire had 20,225, and of females there were 24,764. There are 11 persons to each 100 acres. Hawick had the largest population with 17,445 and Bedrule had the smallest with

180. One hundred and one persons were returned as able to speak Gaelic in addition to English, and none as able to speak Gaelic but not English. Of the persons enumerated 91·2 per cent. were born in Scotland—28,345 were natives of the county. The number of foreigners in both shires was small—22 in Berwickshire, and 49 in Roxburghshire.

In Berwickshire in 1921, of the male population aged 12 years and over, 9569 persons were returned as engaged in some industry or service: and of the female population, 3529. Agriculture employed 4983 workers—4134 men and 849 women, of whom 4520 were directly engaged in farming and stock-rearing. Fishing occupied 587 persons. Personal domestic service required 1863, of whom 1284 were women. Building and contracting required 381 hands: workers in wood and furniture were 245: metal workers numbered 440, but workers in precious metals only 14. The textile industries employed 173 men and 233 women: 184 men and 150 women were engaged in the manufacture of clothing. Skinners and tanners numbered 21: stationers and other workers with paper, including printing and photography, comprised 283: chemical workers numbered 7: 584 persons manufactured food, drink, and tobacco. There were 501 persons engaged in the work of transport and communication, of whom 315 were railway servants: 69 were employed in mining and quarrying. No fewer than 1285 men and women catered for others in various branches of commerce. There were 288 professional men and women—teachers, doctors, clergymen, lawyers, and 744 public servants. Engaged in entertainment and sport there were 41, and in other occupations 22. Out of a total of 22,579 persons of working age no fewer than 982 males and 8499 females had no specified employment.

In Roxburghshire, in 1921, of the male population aged 12 years and over, 14,484 persons were engaged in some industry or service, and of the female population 7734. Agriculture employed 5087 workers—4363 men and 724 women, of whom 4510 were directly engaged in farming and stock-rearing. Personal domestic service required 3307, of whom 2244 were women. Building and contracting required 669 hands: workers in wood and furniture were 378: metal workers numbered 589, but workers in precious metals only 20. The textile industries employed 2320 men and 2922 women: 595 men and women were engaged in tailoring, dressmaking, and bootmaking. Skinners and tanners numbered 117: stationers and other workers with paper, including printers, 92: chemical workers numbered 25: 330 persons manufactured food, drink and tobacco. There were 1122 persons occupied in the business of transport and communication, of whom 878 were railway servants: 50 persons were engaged in quarrying and mining. No fewer than 2531 men and women catered for others as dealers in different departments of commerce. There were 538 professional men and women—teachers, doctors, clergymen, lawyers, and 1266 were occupied in public administration and defence. Engaged in entertainment and sport there were 41, and in other occupations 193. Out of a total of 36,554, no fewer than 1441 males and 12,895 females of working age had no specified employment.

11. Agriculture.

Up to the close of the eighteenth century Agriculture in the Border Counties was not a very successful enterprise. While land and labour were cheap, farmers were poor, and much of the soil lay waste and unused. Rural life was practically untouched by outside influences, and the peasant's lot was, on the whole, a mean and miserable one. Housing was of the worst description, the food coarse, sometimes scarce, the work hard, trying. Nevertheless great advance had been made since the days prior to the Union, when Scottish agriculture was "hardly worthy of the name." The introduction of drill-husbandry into Roxburghshire by Dawson of Frogden gave a new impetus to cultivation, and this was maintained by the invention of the threshing-machine by Meikle, an East Lothian farmer, by Small's swing-plough, and the grain-fanners which were made and used in Roxburghshire before any other county in Scotland. By and by came new and better forms of implements for tilling the soil, including Bell's reaper, the double-furrowed plough, and others. With these, and the introduction of that most potent fertiliser, guano, came the agricultural revival so sorely needed. Many modern improvements have followed, the most novel being that of the petrol-driven tractor.

Thus the whole system of farming has been changed, and what were once wild and uncultivated broad moorlands and spongy swamps, both in Berwickshire and Roxburghshire, have been converted into valuable pasture-lands supporting thousands of sheep and cattle—into fields of waving corn and acres of luxuriant turnip crops.

The general method of farming in the counties embraces stock and crop husbandry. It is the union of these two branches which is the secret of the Border agriculturist's success, and which indeed forms the basis of all advanced and successful farming. In this respect neither of the counties is equalled by any other county in Scotland; for they excel not only on account of uniting the two branches, but in the efficient manner in which both departments are carried out. It is the combination of stock and crop husbandry which has rendered the system of alternating grass and grain crops advisable, and which furnishes the strongest argument for a regular plan of rotation of cropping. The five-course rotation generally practised embraces: (1) Turnips, Potatoes; (2) Barley or Oats; (3) Clover and (4) Grass; (5) Oats. Having two years' grass, the five-course provides for the maintenance of stock, and as it has also a proportionate acreage under turnips, so it raises for stock a supply of winter food equal to its supply of summer food.

According to the return for 1924, the total land area of Berwickshire is 292,535 statute acres; and of Roxburghshire 426,028 acres. The number of Holdings in the two counties is respectively 968 (of which 449 are above 100 acres), and 1168 (of which 471 are above 100 acres). Over the whole of Scotland with a total of 76,210 Holdings, the average size of Holdings is 61·9 acres. In Berwickshire the average size is 195·7 acres, which is the largest in Scotland, while the average in Roxburghshire is 151·7 acres. The largest farms in Scotland belong to Berwickshire, which has 245 above 300 acres, Roxburghshire coming next with 205. These returns do not include hill pasture.

In 1924 the total acreage in Berwickshire under all kinds of crops, bare fallow, and grass, was 189,456—132,576 of arable land and 56,880 of permanent grass. Under wheat there were 1285 acres, under barley 15,680, under oats 28,758, under rye 68, under mixed grain 42, under beans 177, under peas 18,—a total in corn crops of 46,028 acres. Under green crops there were 25,243 acres, which included 21,158 acres of turnips, 128 of mangolds, 50 of cabbage, 963 of rape, 15 of tares or vetches for seed and 332 for fodder, and 2597 of potatoes. There were 61,108 acres under clover and grass in rotation, 56,880 under permanent grass, and 169 bare fallow. Of carrots, flax, etc., there were 11 acres, of strawberries, raspberries, and other small fruits 17 acres. In the same year there were in Berwickshire 4944 horses, 20,680 cattle, 346,664 sheep, and 5611 pigs.

Roxburghshire showed in 1924, 177,160 acres under all kinds of crops, bare fallow and grass, of which 355 acres were under wheat, 9246 under barley, 25,092 under oats, 22 under rye, 5 under mixed grain, 58 under beans, 5 under peas, 1158 under potatoes, 17,482 under turnips, 33 under mangolds, 82 under cabbage, 686 under rape, 8 under tares or vetches for seed, and 122 for fodder. There were 42 acres under carrots and small crops, 18 in bare fallow, 56,716 in clover and grass in rotation, and 66,002 under permanent pasture: $25\frac{1}{2}$ acres were under strawberries, raspberries and other fruits. There were in Roxburghshire in 1924, 4471 horses, 22,180 cattle, 549,219 sheep, and 4095 pigs.

The loamy soil of the shires is peculiarly fitted for barley cultivation. Only two counties in Scotland grow more barley than Berwickshire, namely, Aberdeen and Forfar, which respectively return 18,901 and 17,850 acres. Roxburghshire occupies eighth place in the barley returns, the others, in addition to those mentioned, being East Lothian with 14,923 acres, Fife with 14,101, Kincardine with 9948, and Moray with 9349. Fourteen counties in Scotland grow more oats than Berwickshire, and fifteen more than Roxburghshire. In 1924, of the total produce of wheat in Scotland amounting to 231,000 quarters, Berwickshire produced 6,300 quarters, or a yield of 38·9 bushels to the acre, and Roxburghshire 1700 quarters, a yield of 38·9 to the acre. The barley produce in Scotland in the same year was 683,000 quarters, of which Berwickshire produced 71,000 quarters, or 36·4 bushels per acre, and Roxburghshire 42,000 quarters, or 36·3 per acre. Of oats during 1924 Scotland produced 4,858,000 quarters, Berwickshire's quota being 152,000 quarters, a yield of 42·2 bushels per acre. Roxburghshire's quota was 119,000 quarters, a yield of 37·9 per acre.

Harvest is fairly early in the low-lying parts of both counties. Records show that at West Foulden the average date for commencing harvest is 27th August, and for finishing, 29th September, the elevation being from 200 to 400 feet. These dates may be taken as fairly representative of the Merse, though Tweedside and other isolated localities may average seven to ten days earlier.

The chief stock in our counties is sheep. It is here that some of the most celebrated flocks of Cheviots have their folds. There are several well-known flocks of pure-bred

Leicesters in the district, and on the low-lying farms when turnip food can be got in the spring, Cheviot ewes are crossed with Leicester rams. On the higher grounds "black-faces" abound. By order of the Board of Agriculture all sheep must be dipped twice a year within specified dates. Sheep-shearing is one of the busiest seasons. The fleece is not shorn till the sheep are 16 months old, after that it comes off every July. Sheep-washing among the Cheviots takes place a week prior to shearing-day. The black-faces, however, escape, for as their wool is sold "in the grease" it is said to keep better, and the grease itself is turned into lanoline. Wool-rolling is an art in itself, that of a Cheviot sheep being rolled up with the inside of the fleece outwards, that of a black-face with the wool outwards. The wool of a female sheep is more valued than that of a male.

Although herds of cattle are few in the south-eastern district of Scotland, the Lowland farms have long been celebrated for the excellence of their fat cattle. The breed generally in favour is the Shorthorn, and many capital bulls are annually bred and exhibited at the local shows. But only half the cattle fattened are, as a rule, bred on the farm, the remainder—often crosses—being purchased at fairs and markets. Fattening is pursued in stalls, boxes and courts, but covered courts are now most in favour, not only on account of the benefit to the health of the animals, but also because of the increased value of the manure made therein.

Handsome premiums are paid for first-rate stallions, and the farmers reap the advantage of securing a fine stock of first-class work-horses. A considerable number of thoroughbred and half-bred horses are also produced in the district, and the sporting instincts which characterised Dandie Dinmont continue to be exhibited by the tenant-farmers in the fox-hunting field. The farmers of Berwickshire and Roxburghshire are noted for their skill, energy and indomitable industry, and, as a rule, they are in a contented and highly prosperous condition.

Nowhere are the farm-labourers better paid, better housed, or better educated. In the words of an Agricultural Commissioner: "There is probably no district in Scotland where improvement in cottage accommodation has made so much progress. Indeed I know no county in England where the average cottage accommodation is so good as in Berwickshire—a remark which would also apply to Roxburghshire."

Sales for cattle and sheep are carried on in Berwickshire, at Earlston, Duns, and Reston; and in Roxburghshire, at Newtown St Boswells, Kelso, Hawick, and Newcastleton. St Boswell's Fair, on 18th July, and St James's Fair, at Kelso, on 5th August, are chief events in the Border year.

12. Industries and Manufactures.

Ancient records have references to Hawick wool-merchants receiving safe conduct to go south in quest of business. In 1640 the magistrates of Hawick drew up certain rules for the use of "wabsters" [weavers], and fines were inflicted on those who made linen cloth narrower than an ell and two inches, the width having been fixed by statute.

It was not until the end of the seventeenth century that

the first woollen factories were established in Scotland—at Glasgow and Haddington. A mill at Harcarse and Bogend in Berwickshire employed "many persons both of English and Scots tradesmen, skilful in spinning, weaving, litting [dyeing], waulking [fulling], mixing, and dreeping of woollen yearn and other materials for making of cloath stuff stockings, and others of wool and lint which manufactory is now come to a considerable perfection so that it did make, dreep, and lit as much red cloath as did furnish all the Earl of Hyndforth's Regiment of Dragoons with cloath this last year."

Most of the weaving, however, was done in private houses and was known as "customerwark," payment being made generally in kind—butter, cheese and meal. Spinning was the work of the women, young and old, the "spinsters." The spinster used the "muckle wheel," consisting of a wooden fly-wheel which turned the spindle, and which was set in motion by a vigorous turn of her hand, she retiring slowly backwards, eking out the thread and winding it on the spindle as again she advanced to rotate her wheel. The spun yarn, linen or woollen, was then sent to the weaver who wove it into cloth. By Article 15 of the Union £2000 per annum were set aside for the encouragement and promotion of the manufacturing of coarse wool within those shires which produced that material. In 1729 the Board of Manufacturers sent practical wool-sorters to Hawick, Jedburgh, Lauder, and Galashiels, where the celebrated "Galashiels grey"—a coarse kersey-cloth, 27 inches wide, was the staple product at about two shillings a yard. In 1776 Melrose had as many as 140 looms in operation (none exist now), Hawick had 65, Jedburgh 55,

Hawick

Kelso 40, Lauder 17, Lilliesleaf 14, Yetholm 35, and Earlston 20. Earlston "ginghams" had wide celebrity in their day.

The conception of the Tweed industry in its modern sense began about 1790. A Hawick stocking-maker, William Wilson, a Quaker, who had been in Glasgow for a year or two perfecting his knowledge of the trade, returned to his native place bringing a 20-gauge stocking frame and started business on his own account. In 1798 the weekly wages amounted to only £3. 17s. 3d. In 1804 he took as a partner William Watson, and though the partnership ended in 1819, "Wilsons" and "Watsons" are still the leading firms in the town. Within the century (1820–1920) the manufacture of woollens grew to vast proportions in Hawick. The wildest dreams of its founders could hardly have conceived its vastness. In 1826 the firm of Watson had nine hand-looms: in 1849 they had 95. By 1890 the trade had reached its zenith with 14 firms, 950 looms and 65 spinning sets. There are now only 7 firms, 450 looms and 40 sets, involving a weekly wage-bill equal to about £850,000 per annum.

1500 persons are engaged in the work.

The familiar Tweed originated quite accidentally. In 1826 Messrs Wilson invoiced a consignment of Scottish "tweels" to Locke of London. A clerk misread the word as "Tweed." The name was allowed to stand and was considered the best designation for the particular class of goods made in the Border district.

The extraordinary development of the hosiery manufacture—of more recent origin (though hosiery was made in Hawick by Bailie Hardie in 1790) has limited the Tweed manufacture there. In 1890 there were 1200 stocking-makers and only hand frames, but since the evolution of Cotton's patent frame the output has carried the fame of Hawick hosiery across every sea. There are 18 hosiery firms, with about 400 power-frames at work producing six garments at once, each frame being capable of turning out from 100 to 200 garments per week, depending on the fineness of the fabric.

3500 persons are engaged in this business.

The woollen manufacture began at Jedburgh about 1850. There are now five small mills with less than 100 looms, and about a dozen sets. One mill in Earlston has about 70 looms. The product of the various Tweed factories is confined chiefly to the Cheviot, Saxony and worsted cloths. During the War (1914–18) Government orders for khaki kept every loom working practically night and day. At Cumledge, near Duns, one mill with 100 looms working highest quality of blankets, sends its products everywhere.

Berwickshire was one of the places in which a paper-mill was in operation before the close of the eighteenth century, and there are still two mills in the county, both of long standing. That at Chirnside Bridge, which employs over 300 hands, is the successor of a mill originally seated at Broomhouse, in Edrom parish, and which was reported in 1841 to be on a "very extensive scale." In 1842 a new site was selected for what is known as "Mill 61." This has proved a prosperous concern, and its reputation is high in the home and export markets, the turn-out of paper averaging 120 tons per week. A second and smaller mill is at Bleachfield, on the Eye, near Ayton.

The amount of printing done in Berwickshire is negligible, Berwick-upon-Tweed providing the two county newspapers. On the other hand, at Hawick, Jedburgh, and Kelso, there are large printing establishments, and each town has its own local journals. One of these, the *Kelso Mail*, established in 1797, is the second oldest newspaper in Scotland.

Brewing is practically extinct in the counties. A long-established brewery at Ednam was destroyed by fire in 1885. Melrose brewing is also a thing of the past. A distillery at Eyemouth ceased to function in 1850. Mineral waters are manufactured at Kelso and Hawick. Considerable engineering is carried on. The repairing of farm-implements employs many hands. Carriage building and motor shops give employment to a number of skilled artisans. Several saw-mills are in operation, but meal-milling is now a decaying industry. The picturesqueness of the old mill-wheel has largely disappeared.

Flower and seed-growing, nursery and market-gardening are carried on at Melrose, Kelso, and Hawick. Forestry employs a considerable number of individuals. Both counties possess a number of thriving poultry-farms. Several of these are connected with the Small Holdings System.

13. Mines and Minerals.

Neither Berwickshire nor Roxburghshire is a mining county. Coal exists in both—on the coast of the former, and in Liddesdale in the latter—but to work it is too costly. At Hunthill, near Jedburgh, as shown by the Council

Records, coal was sought for as early as 1660. Attempts made at Mordington and Liddelbank about a century ago met with small success. Peat was in general use, as it is yet in remote districts, and the ancient woods that still remained afforded large quantities of excellent fuel. Now everything is changed, and at the furthest outpost amongst the hills at least one load of coal brought from the nearest station, is set down during the year.

Copper occurs in Longformacus, and upon the farm of Hoardweel, in Bunkle, but repeated digging operations never paid. Antimony is found in Liddesdale. A coarse alabaster or gypsum has been found at Chirnside, Greenlaw, and Kelso. Quicksilver has been got at Holehill, ironstone at Ayton, Mordington, Jedburgh and in Liddesdale. Limestone occurs frequently, and is more plentiful in Roxburghshire than in Berwickshire, as at Carter Fell, Nottylees, Stobs and Lariston, providing also the place-name of Limekiln Edge on the Hermitage watershed. Defunct, desolate kilns may be observed in many places. Great beds of marl exist here and there, one of them giving origin to the name Marlefield, in Morebattle. At one time largely used as manure it has been displaced by more profitable expedients—the nitrates and phosphates that are so common. Some good lapidary stones have been taken from the Tweed. Fragments of agate, jasper, amethyst and rock crystal are occasionally cast up by mole-workings, the plough, and the mountain torrents. Excellent specimens have been found at Roberts Linn, and in the felspar porphyry of the Cheviots.

Brick and tile works are practically extinct in the shires.

There is considerable stone-wealth in the counties, contributing to the amenity of many a town and substantial features in many a country house. Twenty-three quarries are worked by the County Council of Berwick, the average number of men employed being 40. The stone varies in its nature, being mostly red and blue whinstone, and is used entirely for road construction purposes. Excellent freestone of a light yellow colour, used in building, is quarried at Swinton. Bemersyde and Earlston Black Hill quarries yield an admirable quality of red sandstone.

Fifty-five quarries are worked by the County Council of Roxburgh, the number of men employed being 71. In the Melrose district there are 11 quarries, chiefly used for road construction purposes. From a famous quarry at Dryburgh came in all probability the stones for the Roman Camp at Newstead and Melrose Abbey. Thirteen quarries in the Kelso district are nearly all used for road purposes. At Sprouston the monks of old assembled the stones for Kelso Abbey, and out of this quarry Abbotsford was reared. In the Hawick and Jedburgh districts there are 29 quarries employing 33 men. A fine freestone quarry at Belses with stone beautifully grained and of a light-red colour was long a favourite of the builder. The same may be said of Nipknowes quarry, above Hawick, also disused. Singdean quarry, in Liddesdale, is used for the repair of the famous "Knot o' the Gate" pass from Jed Water to the south—the road used by Sir Walter Scott in his Liddesdale "raids," and mentioned in *Guy Mannering*.

14. Fishing.

Fishing is naturally a principal industry of Britain. The shallow seas surrounding our long and irregular coast-line abound in every species of marketable fish. Enormous quantities are used for home consumption, and thousands of barrels are exported annually, especially to the Baltic States. From a variety of causes every branch of the industry has suffered acute depression within recent years. Some of the Continental markets have been closed. Working expenses have risen considerably, and during several seasons, herrings were scarce and in poor condition.

The importance of Scottish fisheries, however, may be gathered from some of the returns for 1921. The total quantity of all kinds of fish (excluding shell-fish) landed in Scotland in that year amounted to 5,260,493 cwts., valued at £4,956,996. Adding the value of shell-fish landed, we have a total of £5,059,328. The total number of persons employed in the three districts into which Scotland is divided, in various branches of the sea-fisheries, was 65,327, of whom 32,183 were fishermen. During the year 535,902 barrels and 153,824 half-barrels were manufactured compared with 1,055,750 full-sized barrels in 1920. In 1921 there were actually belonging to Scotland 391 steam trawlers, 823 steam drifters, 1987 motor boats and 4367 sail boats, a total of 7568, the bulk of these being owned on the East Coast. In 1913 there were 944 more boats of all kinds. The total value of fishing vessels amounted in 1921 to £6,226,818, making with gear valued at £1,680,649, a total of £7,907,467, a decrease of £4,009,001 from the previous year.

Eyemouth is Berwickshire's only seaport. It is an ancient town with a small but excellent harbour opening into a safe, picturesque, semicircular bay admitting vessels of small burden at every stage of the tide. Its main source of employment are the haddock and herring-fisheries, which are largely and energetically followed. Situated on the very key of the coast, access is easy to the best fishing grounds, north and south, and at a moderate distance from land. Railway facilities have greatly helped the progress of the place. The fishermen are of the best type, steady, industrious, enterprising. Their boats are of the most approved class and build, and the fishing materials of the finest description.

About 1840 there were only 10 boats employed at Eyemouth, worth £700, and the estimated value of the fish caught during the year did not exceed £2000. In 1866 34 boats were employed, manned by 215 fishermen, and the value of the fish landed was £15,000. In 1921 the number of sailing boats at Eyemouth was 29, motor boats 81, and steam drifters 23—a total of 133. The total gross tonnage of these was 2674. Their value was £103,834, and with fishing gear £153,522. Of nets there were 5,503,800 yards, valued at £33,865, and of lines 1,329,760 yards, valued at £4732. Crab and lobster creels numbered 2640, valued at £792.

During 1921 573 men and boys were in regular employment. There were 13 fish-curers, 36 coopers, 247 gutters and packers, 74 carters and labourers, 70 persons gathering bait and baiting lines, 15 boat builders, 12 persons

Eyemouth

making and mending nets, and 77 engaged in other occupations—a total of 1117.

The Fishery Board for Scotland controls all matters
relating to the fishing industry. Under Eyemouth District
are included 16 English stations on the Northumberland
coast, with the creeks of Burnmouth and St Abbs. Cove
is in the Leith District.

Fisher girls at work, Eyemouth

Fish are classified as *demersal* and *pelagic*, that is, those
which live and feed mainly near the sea bottom in fairly
shallow waters, and those which live near the surface. The
demersal, again, is subdivided into (1) *round fish*, e.g. cod,
ling, torsk (or tusk), saithe (coal fish), haddock, whiting,
conger eels, gurnards, catfish, monks (anglers) and hake;
and (2) into *flat fish*, e.g. turbot, halibut, lemon sole,

flounders, plaice, brill, dabs, whitches, megrims, skates, rays and squids. These are caught chiefly by line and trawl. Pelagic fish are herring, sprats, sparling and mackerel, and are caught mostly by drift-net, though some of the Eyemouth fishermen, in common with those of Berwick-upon-Tweed, use what is called the Danish seine-net or snurreraad which, being shot in a semi-circle, has its ends drawn together, whereas in the drift-net the herrings are held in the meshes by their gills.

The total quantity and value of all kinds of fish landed at Eyemouth during 1921 was 63,392 cwts. valued at £44,215, a decrease on the previous year of 8812 cwts. and £17,339. During 1921, 36,011 lobsters valued at £2628, and 678,800 crabs valued at £7447, with 336 cwts. of unclassified shellfish, valued at £74, were taken, an increase on the previous year of £1378. The number of barrels cured was 29,114, and the total value of cured fish was £119,905.

As with all East Coast stations, the Eyemouth boats participate in the East Anglian herring fishing, at Yarmouth and Lowestoft. Forty-six crews went south in 1921 where their gross total earnings amounted to £16,328 as compared with £61,030 for 42 vessels in 1920. The fishing proved unprofitable and "a number of crews found themselves in debt at the close of the season."

The salmon fishings in the Tweed are valuable. In 1807 the annual rental amounted to £15,766, and the number of boxes of salmon sent to London that year was 8445, each box containing 8 stone weight. In 1858 the rental had fallen to £4000. The Tweed Assessment Roll for 1923 shows that in Berwickshire 21 proprietors shared the different waters, including tributaries, and 11 in Roxburghshire. The total rental in 1923 of the whole Tweed watershed was £15,196. 19s. 7d.

The Duke of Roxburghe owns the most extensive fishings in the river—from Sprouston to Rutherford, a stretch of about 10 miles. Some of the finest water is that at Birgham, Hendersyde, Floors, Makerstoun, Mertoun, Dryburgh, Bemersyde, Old Melrose and Pavilion. The average weight of salmon caught in the Tweed may be stated at 9 or 10 pounds, but salmon are frequently captured with the fly on the Kelso waters weighing 40 pounds, the largest record being a male fish of 57½ pounds, caught by rod at Floors in 1886. Another which weighed 60 pounds was found dead in Mertoun Water in 1907. Grilse vary from three to seven pounds. The annual close time for net-fishing on the Tweed is from 15th September to 14th February. For rod-fishing it is from 1st December to 31st January—giving the longest open time in Scotland. The total weight of salmon carried by rail in 1921 was 2731 tons, of which 1736 tons were caught in the area from Berwick to Cairnbulg Point, at the entrance of the Moray Firth. Of this a considerable proportion must have been contributed by the salmon fishings of the Tweed. The Tweed District is under the management of Commissioners who meet periodically at Kelso.

Trout is abundant in every river and stream. Pike are frequently taken in the lower reaches of the Tweed. A 35-pounder was taken in Hirsel Loch in 1837. Roach made their appearance in the lower Tweed in 1898. The first to be caught in Teviot was in 1910. Small dace have been seen. Grayling are common in the Teviot. Perch are found in Coldingham and other lochs. Gudgeon is plentiful in the Leet at Coldstream. Minnow ("mennant") and stickleback ("baneytickle") are common. Loach ("bairdies") exist in every burn. An occasional tench is netted in the Tweed. Eels are abundant, following most of the streams to their sources. Sea-lamprey have been taken as high up as Kelso. The mud lamprey ("ramper") is common. Vast numbers of salmon ascend the Tweed and its tributaries for spawning from October to December.

15. History of the Counties.

About A.D. 80 the Roman general Agricola marched his legions north of the Cheviots, through modern Roxburghshire or modern Berwickshire. From the time he took to reach the Forth, we gather that the difficult ground and the warlike inhabitants made campaigning far from easy. Our counties became part of the Roman Empire for some time, and for longer were influenced by intercourse with the Romans; but no trustworthy details can be given.

Following the exit of the legionaries in 410, the Border district became an arena of constant warfare between the Picts and Scots, and the Britons, until the sixth century, when it reappears in history as the Anglian kingdom of Bernicia. This is the epoch associated with the victories of Arthur, shadowy leader of the Christian Britons. Arthur died about 537. Popular fancy regards him as merely asleep within "Eildon's caverns vast," awaiting the horn which is to peal his march from Fairyland.

In the next century, Bernicia re-united with Deira

(modern Yorkshire) to form the powerful realm of North-umbria, extending from the Humber to the Forth. For the next three or four hundred years the story of the Border is little more than a wild record of lawlessness and blood-shed. It had grown into a sort of battle-ground for every hostile tribe from far as well as near—the Danes, for instance, the latest of the invading hordes. From a monarchy Northumbria fell to the level of an earldom in 954. In 1018 the Scots crushed the Angles of Northumbria in a notable victory at Carham-on-Tweed, and won the territory known as Lothian—the country between Tweed and Forth. Thus at the dawn of the eleventh century we have the Tweed constituting the virtual as it was the natural boundary between the two countries, though not until 1222 was the actual line of demarcation—the Tweed, the Cheviots and the Solway—adjusted by an international Commission.

We do not know when the light of the Christian faith first penetrated the Border country. Paulinus, whom Edwin's Kentish queen brought to York, may have been its earliest evangelist. The conversion of Northumbria was completed during the reign of Oswald, friend of Aidan and Boisil, founders of Old Melrose. St Cuthbert was the great Apostle of Tweedside, carrying the gospel to and planting churches and chapels in the remotest parts. Queen Margaret, wife of Malcolm Canmore, had her share as an ameliorative and spiritually and intellectually enlivening influence. With the accession of her son David I, the Border emerged not only into distinct existence, but it assumed an importance unattained by any other part of the country. The towns of Roxburgh, Jedburgh, Kelso, came into prominence about this time—Roxburgh being a favourite royal residence. The establishment of the four Border Abbeys may be described as the crowning glory of what was the most peaceful and progressive era in Scottish mediæval history. Under Alexander III, Border life and literature continued to flourish. At Jedburgh Alexander wedded a second time. While the hilarity of the marriage-feast was at its height, an unwelcome visitor—a skeleton figure of Death—appeared. It was no doubt a mere *bal masqué* effect, but a superstitious age read into it an evil omen—a feeling which was substantiated by Alexander's tragic end a few months later.

With the passing of Alexander and the death of the Maid of Norway, whose betrothal to Prince Edward of England had been ratified by the Scottish Estates at Birgham in 1290, the period of Blood and Iron on the Borders began. Edward I, despotic and ambitious, sought the annexation of Scotland, and his suzerainty was recognised at a great assembly at Norham in June 1291. As the Ragman Roll shows, practically every Border family of note did fealty to Edward at Berwick, then the most important seaport in Scotland, and spoken of as the "Alexandria of the North." Next year John Baliol was adjudged rightful heir to the crown. Edward remained *de facto* ruler, and Baliol did him homage. But matters so eventuated that Scottish patriotism was roused to its full height. At Bannockburn in 1314 Robert Bruce became master of Scotland. In 1318 Berwick fell to Bruce. In 1322 English invaders retreating from Edinburgh, set fire to the abbeys of Dryburgh and Melrose. In 1329 Bruce died. His heart, which he wished buried in the Holy Sepulchre at Jerusalem, was brought back to Melrose after the Good Sir James Douglas, courageously carrying out his commission, died fighting with the Saracens in Spain. On 19th July, 1333, was fought the battle of Halidon Hill, a steep eminence overhanging Berwick, where the Scots suffered severe defeat. Berwick's surrender closed its career as a Scottish town. In 1388 a Scottish army, 50,000 strong, assembled at Southdean, and in the moonlight of the 19th August, Otterburn, "the most spirited fight in history," as Froissart styles it, ended with disaster and with glory. "Langsyne I heard a prophecy that a dead man should win a field, and I trust in God that dead man shall be I," muttered the brave Earl of Douglas in his death-throes. Ere morning broke Hotspur had been captured, and the routed, decimated English army was in full flight. In Melrose Abbey rests the dust of the Border hero.

The capture and demolition of Jedburgh Castle by the Scots in 1409, their seizure of Fast Castle the following year, the siege of Roxburgh Castle in 1460 (costing the life of James II) and its subsequent destruction as a pre-cautionary measure against further assaults, the annihilation of the town of Roxburgh: the coronation of James III in Kelso Abbey: and the hanging of that monarch's favourites across the parapet of Lauder Bridge in 1482, are chief events in our history during the fifteenth century.

On 9th September, 1513, within sight of the Tweed, James IV and the flower of his army fell on Flodden Ridge. In 1514, tradition tells how a company of raiders threatening the town of Hawick, found their match at Hornshole, where a party of the younger townsfolk who had little experience in arms, overcame them and bore

Melrose Abbey

back in triumph an English pennon. In 1523 Lord Surrey burned the town of Jedburgh. In 1526 occurred near Melrose the last great clan-battle of the Border, when

> Gallant Cessford's heart blood dear
> Reeked on dark Elliot's Border spear,

the origin, says Scott, "of a deadly feud betwixt the names of Scott and Ker which cost much blood upon the Marches." In 1529 James V ordered the execution at Carlanrig, on the Teviot, of Johnie Armstrong and twenty-four of his followers—a callous deed of misdirected vengeance. In 1542 Sir Robert Bowes and the Earl of Angus with 3000 horsemen were on their way to harry Jedburgh but were ignominiously routed at Hadden Rig by the Earl of Huntly. As a retaliation 30,000 English marched into Scotland under the Duke of Norfolk and burned the town and abbey of Kelso.

The ill-fated Queen of Scots paid several visits to the Border. She was at Jedburgh in 1566 holding a Court of Justice. Her lover, James Hepburn, Earl of Bothwell, lay wounded in his castle of Hermitage, the result of an encounter with "little Jock Elliot" of Park. Mary heard the news and probably imagining that Hermitage was nearer than it was, set out on what was a rough journey of 25 miles by hill and moor. She remained only an hour or two, "to Bothwell's great pleasure and comfort," riding back to Jedburgh the same day, to fall ill of a virulent fever which well-nigh finished her career. Later years are said to have evoked the reflection, "Would to God I *had* died at Jedburgh!"

Berwickshire and Roxburghshire had their share in the Covenanting episode. Duns Law was the venue of General

Leslie's troops in 1639. It was at Coldstream that General
Monk mustered the army with which he marched south
to bring about the Restoration of Monarchy in 1660.
Monk's regiment became the Coldstream Guards.

With the Revolution of 1688 came peace and religious
liberty to the Border. The Union of 1707 brought its own
beneficial results. When Queen Anne died in 1714 the
question of succession made the attempt of 1715 possible.
But it was a half-hearted affair so far as the Border was
concerned, though King James was proclaimed in the
market-place of Kelso with drums beating, banners waving,
and bagpipes playing.

The Rising of 1745 met with similar indifference.

Over the advent of the nineteenth century hung the
fear of Bonaparte's threatened invasion. In the earlier
centuries a system of communicating intelligence by means
of a bale-fire blazing from the hill-tops and bartizans of
the peels had been common. This was revived. All who
could bear arms enrolled in the Yeomanry and Volunteer
companies which were formed in every parish. On the
evening of 31st January, 1804, the watch at Hume Castle
observing a light flaming afar off mistook it for the long-
expected signal. They fired their beacon. Very soon all
through Teviotdale, Tweeddale and Liddesdale, the glow-
ing summons was repeated from point to point. Bustle
and confusion were everywhere. The market-places were
filled with men on parade. Many rode from long distances
and the old Border spirit shone as resolutely as ever it had
done in stormier days. Even if it was a False Alarm it told
its own tale of duty and of a chivalry and patriotism which
men of the Border breed have never been slow to exhibit.

Infinitely more must be said of the happenings in the Great War (1914–1918). Its unexampled record saddened and illumined every hamlet throughout Merse and Teviotdale with unforgettable grief and glory.

16. Antiquities.

The earliest implements found in Britain belong to the period when our island was still part of the Continent, and we can tell little or nothing as to their age. This is the pre-historic epoch which antiquaries, having no other means of dating, have divided according to the material of which man formed his implements, into the Stone, the Bronze and the Iron Ages, the first of these being sub-divided into the Palæolithic or Old Stone Age, and the Neolithic or New Stone Age (probably to about 1800 B.C.). Of Palæolithic man there is believed to be no trace in Scotland, but of the presence of his successor, the Neolithic Iberian who improved on the chipped weapons of his predecessor, and learned to grind and polish them, there is ample evidence.

Arrow-heads—leaf-shaped, barbed, and lop-sided—scrapers used in the dressing of skins for clothing, saws, borers, and knives of flint have been recovered in considerable numbers in the Lauderdale uplands and on farms near Earlston as well as at Jedburgh, Harden, Bowden and Melrose. Hammers which were held in the hand have been frequently turned up by the plough. The finest axe yet seen in Scotland of green quartz, $9\frac{3}{4}$ inches in length, $4\frac{1}{8}$ inches across the cutting edge, and $\frac{9}{10}$ of an inch in thickness, was found in Berwickshire about the year 1840. Stone axes have been found at Marlefield, Edgerston, and at Bloodylaws, in Oxnam Water. A perforated hammer of gneiss was picked up at Bonchester Bridge, in Rule Water. Whetstones, whorls (used in spinning), stoneballs, discs, and tools of

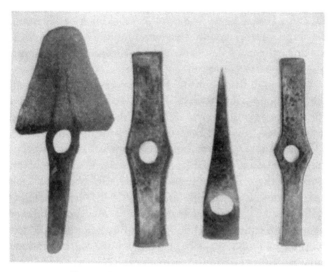

Roman implements found at Newstead

various shapes, some of them for unknown purposes, have been collected all over the counties.

During the Stone Age man discovered the use of metal. He had not skill to smelt the iron, but he was able to bring together the elements of copper and tin, and to introduce the Age of Bronze (probably 1800 to 500 B.C.). Remains of a broad-bladed lance or spear-head of bronze were unearthed at Craigsford Mains. Flanged axes of this material have been found at Windshiel, Duns, Clinthill, Longcroft, Gattonside, and at Sudhope, near Jedburgh. A ferrule of bronze for a spear-shaft was picked up at Whitsome, and is the second example of its kind known to have been found in Scotland. England can show a greater number. In Corsbie Moss, Legerwood, and at Cockburnspath bronze swords in perfect condition were found. A javelin head was unearthed at Ellemford. At Southdean and Dryburgh axes of copper were found.

The men of the Stone and Bronze Ages lived very much like the modern Hottentots, in kraals of willow and mud. For places of refuge and defence they had underground earth-houses, of which Berwickshire furnishes one example, at Broomhouse, Edrom, and crannogs or lake-dwellings constructed on artificial or partly artificial islands on lakes. Probably the only crannog known to the Border district is a curious construction found in Whitburn Moss, at Spottiswood. Uncivilised as they were, the people of those remote ages entertained a due respect for their dead. Bodies were buried intact, or were partially cremated, the calcined remains being placed in urns, over which was piled a huge cairn of stones. No fewer than 465 burial sites of the Bronze Age have been recorded for Berwickshire. Lauder heads the list with 115, followed by Cockburnspath with 84, and Coldingham with 73. The number of cairns on record is 341. Of those still *in situ*, mostly unexcavated, there are 216, 36 being over 20 feet in diameter.

The Mutiny, Mittenfull, Meeting or Mitten o' Stanes (Armstrong's Map 1771), on Byrecleuch Ridge, a heathery moor in Longformacus, 1250 feet above sea-level, is the only example of its class in the south-east of Scotland. Lying with its longest axis east and west it measures 278 feet in extreme length, 26 feet in breadth at the west end, suddenly expanding at 278 feet eastward, and showing a frontage of 76 feet at its eastern extremity. It is only three feet high at the west end, but gradually rises till, at the centre of the east end, it has an elevation of 11½ feet. A large sheep-fold has been built out of it along the south side, and otherwise it has suffered much from dilapidation. Recent excavations have discovered nothing of note. The Twinlaw Cairns, in Westruther parish, conspicuous landmarks of the Lammermoors, were excavated and reconstructed some time ago. A stone cist, previously disturbed, was found in each, and drawings were made by Lady John Scott. At Clacharie, in the parish of Lauder, an urn contained charred bones, and six cists with unburned bones but no urns. Two cairns excavated in the Hagg Wood at Foulden disclosed several cists and a number of interesting relics. The number of cists recorded in the county is 180, and of urns 67.

In Roxburghshire, tumuli exist or existed at Langraw, on Rule Water, Fodderlie, Southdean, Linton, Morebattle, Caverton Edge, Crailinghall, Eckford, Wooden, and Yetholm. Hawick Moat or Mote is the most interesting and perfect specimen of these ancient burial mounds. It has never been investigated. It is conical in shape, flat on the top, measuring 30 feet in vertical height, 117 feet in circumference at the summit, and 312 at the base, and has a cubical content of 4060 yards.

Stone-circles of unhewn stone are thought to be the work of Neolithic people, and those of hewn stone the work of the Bronze Period. But nothing can be said with definiteness. They have been described as Druidical temples, or Druidical halls of judgment—a cult of which nothing authentic is known. Whether they were places of worship, or, as some think, mere clocks to mark the speed of time by the passage of the celestial bodies, may be solved some day. Meantime all theories are pure guess-work. Excavation has proved that they were places of sepulture, as no doubt were also the numerous standing-stones throughout the country. Berwickshire has only one specimen of a stone-circle, at Borrowstoun Rig, in Lauder-dale, where 32 unusually small stones are *in situ* with a recumbent centrical slab termed the "altar-stone," prob-ably a misnomer. The diameter of the entire circle is 140 feet. These circles greatly vary, and the centre-stone is frequently missing. Roxburghshire has eight well-defined stone-circles. The best-known is that on the Nine-Stane Rig, in Liddesdale, of which only seven stones are identi-fiable. It was here that Lord Soulis was boiled to death for his crimes, according to Leyden's ballad, which, however, rests on no historic basis:

> On a circle of stones they placed the pot,
>   On a circle of stones but barely nine:
> They heated it red and fiery hot,
>   Till the burnished brass did glimmer and shine.
>
> They rolled him up in a sheet of lead,
>   A sheet of lead for a funeral pall;
> They plunged him in the cauldron red,
>   And melted him, lead, and bones, and all.

Other circles are at Tinnis Hill, on the Dumfriesshire boundary; Brugh Hill, near Hawick; Stonedge; Trestle Cairn Townhead Hill, and Frogden.

The Brethren Stanes

The Standing Stone near Earlston

Several single stones are scattered over both Berwickshire and Roxburghshire. One, known as "the Standing
Stone," is on the road to Morriston, near Earlston. At
Brotherstone two greenstone uprights are locally known
as the "Brethren Stanes," from a constantly repeated but
unsubstantiated tradition. Another is at Thirlestane.
A fourth is the Pech Stane at Billie Mains. In Roxburghshire the most remarkable of this type is the "eleven
Shearers of Hownam," set in a straight row and said to
represent unhappy individuals who dared to shear their
harvest on a Sabbath. Only one cup-marked stone has been
noted—at Chirnside. It is an interesting specimen.

Berwickshire is rich in defensive constructions. No
fewer than 92 have been recorded, all of native origin.
Many more have been swept out of recognition by the
march of agriculture and the ruthless quarryman. A group
of ten are located within a radius of a mile on the high
rough moorland around Coldingham Loch, an admirably
fortifiable site. In Lauderdale this type of fort is numerous
—fourteen being in Lauder parish alone, and six in Channelkirk. Along Bunkle Edge, which commands a wide prospect across the Merse and into England, in a distance of
less than two miles from east to west six defences and sites
can be counted. Fifteen others are traceable more or less
in Cockburnspath. It is interesting to remark on the
character and situation of these prehistoric defences. They
consist of (1) *cliff* or *escarpment forts*, depending in parts
on a cliff or on a steep slope of defence—either on the sea-
shore or inland—of which Earnsheugh overlooks the
North Sea, and Ninewells overlooks the Whitadder.
Heugh, above Blythe Water, Milne Graden, and Lennel

Hill are also good examples; (2) *promontory forts,* in which the main fortification necessary is that across the base or neck of the hillside projection, as at Raecleughhead in Langton, and Wallace's Knowe in Lauder parish; (3) *contour forts,* which, again, are divided into (*a*) *circular forts,* e.g. Tollis Hill and Thirlestane Hill, (*b*) *oval forts,* e.g. West Addinston and Habchester, and (*c*) *rectangular forts,* of which only one example at Bunkle Edge has been noted; (4) *small enclosures,* more domestic than defensive in character and in which hut circles abound, as in the remarkable group around St Abbs Head and Coldingham Loch; (5) *brochs,* e.g. that at Edinshall on Cockburn Law, overlooking the Whitadder, an imposing monument of unexcelled dry-stone masonry in height about five feet; (6) *motes,* the mound on which the early palisaded wooden tower was erected, of which the only example in Berwickshire is The Mount at Castle Law, Coldstream. In Roxburghshire the Mote at Hawick has attained wide celebrity; (7) *large enclosures,* partially excavated in the interior and intended as folds for live stock.

In Roxburghshire forts exist or existed on the north side of the Tweed, between Gattonside and Leaderfoot, at Camp Knowe, Chester Knowe, and Easter Hill; on the south side of the river, at Cauldshiels, Haxil Cleuch, Marslee, Kaeside, and Eildon, the largest fort in the county, girdled by a triple line of defence about a mile in compass; at Ringley Hall, near Makerstoun; in Teviotdale, at Penielheugh and the Dunian; on the Kale at Hownam Law; on the Liddel at Caerby, Tinnis, Needslaw, Hudhouse, Flight, Sorbietrees; on the Hermitage at Shaws, Tofthole, Foulshiels, Blackburn and Cocklaw.

No definitely Roman construction is found in Berwickshire. In Roxburghshire the great Camp at Newstead is the outstanding relic of the Roman occupation of the

Brass Helmet (Roman) found at Newstead

Border. Situated in a field commanding spacious views of the Tweed valley between Leaderfoot and Newstead, this Camp underwent a thorough exploration in 1905—10.

The work has been finely described by Dr James Curle in
A Roman Frontier Post and its People. Newstead is the
largest of the Roman camps excavated in Scotland, covering
an area of 20·824 acres. The fort was of the usual rectangular
shape, greater in length than breadth, measuring 810 feet
by 720 feet. The corners were rounded, and the four gates,
one on each side, were placed opposite each other. The
defences consisted of three parallel lines of ditches, a wall,
and a rampart. Among "finds" were six altars dedicated
to different deities; a large number of querns; bronzes;
brasses; ornamented leather work; helmets; armour;
swords; spears; harness; tools and implements for various
purposes; lamps; weights; coins; seals; beads; writing instruments; and an immense quantity of pottery ware. Most
of these have been deposited in the Museum of the Society
of Antiquaries at Edinburgh. The Roman camp at Cappuck
on the Oxnam was partially excavated by the Marquess of
Lothian in 1886.

Caves artificially hewn out of the Old Red Sandstone
cliffs of the Jed exist at Mossburnford (1), Lintalee (1),
Hundalee (1); of the Ale, at Ancrum (18); of the Oxnam,
at Crailing (8); of the Kale, at Grahamslaw (9), and of
the Teviot, at Sunlaws (5)—43 in all. It is impossible to
say what their object was or when they were constructed.
All are of considerable dimensions, and access must have
been a formidable undertaking.

At one time Crosses were found in every corner of the
land. They marked boundaries, battle-scenes, murders, and
important events. The "mercat-cross" was not merely a
commercial centre. Proclamations were made from it, and
it was a place for prayer and preaching. Few of the ancient

Crosses of the Border are extant. Those of Cockburnspath, Coldingham, Greenlaw, Ancrum, Bowden, Maxton and Melrose still exist, though in some cases restored. Crosses at Crosshall in Eccles, and Milnholm on the Liddel, are interesting relics.

Many ancient coins have been found in the Border area—coins of the Roman epoch, others struck in the great mints at Berwick and Roxburgh, bearing the legends "Eola on Berv." and "Hugo on Roch," being the name of the mints and the place of coinage, Roch standing for Roxburgh. Coins of the Edwardian era—Edward I, II, III—were found at Jedburgh on the demolition of its castle. In 1827 a hoard of 90 silver coins was dug up in a field at Bongate, Jedburgh, one of which was of Canute's reign. In 1836 a child playing at Hayhope near Yetholm found a large collection of early British coins. At Earlston, in 1787, two horns were found containing gold and silver and copper coins of the Jameses and Queen Mary. At Greenlaw, Eccles, Coldingham, Duns, similar hoards were found, that at Duns numbering 2361 pennies of Edward II, Alexander III and Robert I.

The most puzzling relic of Border antiquity is what is known as the Catrail. It consists of a ditch with a mound running along each side, of varying depth and width, and extending in an erratic semicircular direction showing considerable breaks and blanks here and there from Peel Fell among the Liddel Water Hills to Torwoodlee on the Gala—a distance of over 50 miles. Antiquaries are much divided as to its object. Some consider it to have been a work of defence, and see in its name a proof of their contention—*cat* or *cad*-rail being supposed to mean "the

Cross at Crosshall, Eccles

battle, or dividing fence." Others believe it to have been
a road for conveying cattle unseen by the enemy. That,
however, is hardly likely. Others, again, maintain that it
was a boundary line. For this there is much to be said.
The latest investigator (1924) forms the conclusion that
the Catrail must be abandoned as a myth so far as the
conception of a continuous united line is concerned. He
puts in its place two portions of black-dyke only—the
True Catrail, from Robert's Linn to the Hoscote Burn
($13\frac{1}{2}$ miles), and the Picts' Work Ditch, from Linglie
Hill to Mossilee near Galashiels ($4\frac{1}{4}$ miles)—the deepest
and most important, if not the longest, of many Black-
Dykes in the Border country. (See Map.)

17. Architecture (*a*) Ecclesiastical.

The earliest Scottish churches were simple constructions
of earth and wattle. In course of time walls of oak and
pine replaced those temporary expedients. These in turn
were succeeded by structures with a more solid foundation
on which uprose masses of strong rubble-work, the whole
being covered in by a roof of rushes or heather. The floor
was of clay, the door narrow and low, and a single window
lighted the gloomy vault-like apartment. A better epoch
followed in the twelfth century with the introduction of
the Norman or Romanesque style. Though the buildings
of that age were characterised by a severe simplicity the
carver's art began to show itself in an ornamentation of
portions of the rounded and circular arches and doorways
which were main features of the period. The first half of

Map of the Catrail

By J. Hewat Craw

the thirteenth century introduced the First Pointed or Early Gothic Period, in which generally the pointed arch took the place of its rounded predecessor. Attention was centred on ornamentation and Nature became the chief prototype. This was quickly followed by the Middle Pointed or Decorated Period in the fourteenth and fifteenth centuries, and this, again, by a gentle gradation passed into the Third or Late Pointed or Perpendicular Period. The larger churches of this period are nearly all restorations, and only churches on a smaller scale were erected. These consist, for the most part, of a single compartment without aisles. The east end frequently terminates with a three-sided apse, a characteristic borrowed from the French architects when the "Auld Alliance" was at its height, a distinguishing feature, the pointed barrel vault, being almost universally employed. The windows were low; the door-ways of the old round-headed form, with late foliage and en-richments. Porches were occasionally introduced. Towers were common, but in many instances remained unfinished or were completed with short spires containing small dormer windows. Monuments and coats of arms set in arched and canopied recesses were of frequent occurrence, and much of their rich carving may have come from the hands of French masons who flocked to Scotland during the reigns of James IV and James V.

The Abbeys are the noblest examples of ecclesiastical architecture in the Border counties. Jedburgh is the oldest. It was founded as a priory about the year 1118, and raised to the dignity of a monastery about the end of the "soir sanct's" reign. Dedicated to the Virgin, it was endowed with vast wealth, and was tenanted by a colony of

Jedburgh Abbey

Augustinian friars from Beauvais in France. Even as a ruin Jedburgh is noble and majestic—a veritable poem in nature and art. Frequently battered and burned, worst of all by Surrey in 1523, and by Evers and Hertford in 1544–45, there are evidences of at least four restorations. It is chiefly of the Norman and Early Decorated Period, and its west and south doorways are among the finest extant specimens of the architectural art. In 1559 the monastery was suppressed and left a roofless ruin. By and by the nave was transformed into the parish church and used as a place of worship till 1875 when a new church was erected upon another site.

Kelso Abbey, founded in 1128, was not completed till well on in the next century. As Earl of Huntingdon, David had planted a colony of thirteen Benedictine monks at Selkirk. After his accession, considering Selkirk unsuitable for an abbey, he transferred them to "the place called Calkou" by the Tweed. This abbey is held to have been the finest example in the country of the castellated style of ecclesiastical architecture. It was a massive graceful pile of grey stone, in form a double cross, with three doorways, three naves, divided by a double row of columns. It had two towers, one by the entrance to the church, the other in the interior part at the choir, squared in plan and crowned by pyramidal roofs like the Basilica of St Peter's at Rome. The discovery of a document in the Vatican Library sheds new light on the whole structure and solves what was a complicated problem as to its original appearance. Generally it is late Norman, with here and there a touch of Early Pointed work. None of the Border abbeys suffered as Kelso did from devastating armies. From the times of Bruce and Baliol to its final destruction by the English under Hertford in 1545, the abbey was frequently laid waste. After the Reformation a low rude vault con-

Kelso Abbey

structed over the transept served as the parish church until 1771, when, during worship, a piece of plaster detached from the roof, fell among the congregation, who hastened out in terror believing a reputed prophecy of Thomas the Rhymer that "the Kirk would fall when at its fullest" was about to be fulfilled. The deserted building was never again occupied.

Dryburgh Abbey

Dryburgh Abbey, in Berwickshire, is perhaps the most charmingly situated monastic ruin in Britain. Dating from 1250, the honour of its foundation is usually ascribed to Hugo de Morville of Lauderdale, and his wife, Beatrix de

Beauchamp. Though David I spoke of himself as "the founder," the balance of evidence favours the Morvilles. It was peopled with a chapter of Premonstratensian monks (white canons) from Alnwick, and was at the zenith of its glory for a comparatively brief period, between 1296, when the canons took the oath of fidelity to Edward I, and 1322, when Edward II burned and pillaged it, provoked, it is said, by the imprudent monks having rung a triumphant peal on the occasion of his retreat after his unsuccessful invasion. Never fully restored to its pristine strength and beauty, Dryburgh suffered the fate of its sister fanes from hostile English attacks till its final burning by Hertford in 1545. The building, of local sandstone, was of the usual cruciform shape. Choir and nave are 190 feet long, and the transepts extend to 75 feet. The nave with aisles is 55 feet wide, and the cloisters a magnificent square of 100 feet. Cellars and refectory, the latter containing a superb rose window of twelve divisions, measure 100 feet by 30. Saxon, Transition-Norman (as in the finely-moulded west doorway) and English Pointed (as in St Mary's aisle) are the prevailing architectural styles. Under the arched groined roof of St Mary's aisle rest the dust of Sir Walter Scott, his wife, his eldest son, and son-in-law—a shrine for thousands of pilgrim feet from every land.

Melrose is by far the most celebrated of the Border abbeys. As an architectural ornament it is possibly the grandest specimen of Gothic extant. Built from designs by John Moreau, a Parisian architect, between 1130 and 1146, it became the abode of the first Cistercian colony in Scotland. But it, too, suffered considerably from Edward's army in 1322, and again (after a noble reconstruction),

Melrose Abbey

from Evers and Latoun in 1544. Poetic justice awaited them, for soon afterwards they were defeated and slain at Ancrum Moor. In 1326 Robert the Bruce gave £2000 (£60,000 at to-day's reckoning) for its rebuilding. The ruin as it now stands, however, is of later date—in red sandstone of the most durable quality which still retains a wonderful sharpness notwithstanding the weathering of centuries. The architecture itself is a mixture of Second Pointed, Perpendicular and Flamboyant. At the Reformation it was finally dismantled, and for long years the ruin was used as a quarry by the Melrosians, some of its stones being still distinguishable in local residences. Up to 1810 the nave was used as the parish church.

The four Border abbeys have been placed under His Majesty's Office of Works, and extensive reparations are in progress upon each.

The remains of three other monastic establishments belong to Berwickshire. The Priory of Coldingham, founded in 1089 by Edgar, King of Scots, was bestowed on the monks of St Cuthbert at Durham and endowed with valuable possessions in the Merse. Extant portions comprise the choir patched up for use as the parish church, the walls of the south transept, and several pillar-bases— those which supported the central tower. Excavations now being carried on have exposed the sites of the chapter-house and of the cemetery. The unique position of the cloisters has also been determined—the interesting fact being revealed that they adjoined the choir on the south side, and not the nave as erroneously shown on old plans. Of the Cistercian Nunnery at Eccles a fragment only is left. Abbey St Bathans church contains a portion of its ancient Priory.

The interest of Ladykirk church lies not only in the fact of its erection in 1510 by James IV in gratitude for his rescue from possible drowning in the Tweed, but also because it is one of the last of the Scottish pre-Reformation churches still quite entire.   It is a perfect specimen of

Ladykirk Church

the architecture of the period—cruciform, with apsidal terminations to nave and transepts, and a tower at the west end. The roof is of stone and the whole building is heavily buttressed. A round tower in Cockburnspath church is the only example of its kind south of the Forth.

Of the ancient parish churches of Roxburghshire interest centres in the following. Smailholm is distinctly a Norman structure, although it was considerably altered in the seventeenth century and at a recent date. It was one of the churches of Scott's boyhood and contains a window memorialising his stay at Sandyknowe. The church at Linton (restored) pertains to the Norman period. It possesses a Norman font, and a sculpture on the tympanum of the

Laird's Loft in Bowden Church

ancient doorway is said to represent the slaying of the legendary Worm of Wormiston by Sir John Somerville. Though it is a constant tradition of the district, and the monster's "den" is still pointed out, the story is probably only a local adaptation of St George and the Dragon, or the tympanum may merely carry an allegorical conception of Faith triumphing over Evil. St Mary's at Hawick dates as far back as 1214. Once a very ornate edifice not unlike a Scottish abbey on a smaller scale, all its ancient glory has departed, and instead of continuing to be the kirk of the parish, it has now only the status of a *quoad sacra* building. Of Hassendean, one of the most important ecclesiastical buildings in Teviotdale, not a vestige remains. Belonging to the Transitional-Norman period, it must have been a beautiful structure, to judge from an etching (1788) in Cardonel's *Picturesque Antiquities,* our only sketch of the ruin. Bowden church is one of the oldest in Scotland and its pre-Reformation nave, chancel, and north transept remain almost intact. Within the church and opposite the pulpit is a perfect specimen of a "Laird's Loft." St Boswells church has some relics of carved corbels and other fragments of Norman date. The western doorway of Maxton church furnishes a fair specimen of Norman style, but the church itself is comparatively modern.   Up to 1791 it had a roof of broom.

18. Architecture—(*b*) Castellated.

Though the Normans introduced castle-building upon a spacious scale into England, their influence only sparsely extended to Scotland. No distinctively Norman military structure exists north of the Cheviots. The castellated buildings of Berwickshire and Roxburghshire range in date from the fourteenth to the 17th century. Those of the thirteenth century, with which the First Pointed Period is associated, have perished, except, perhaps, the wall-foundations of Hume Castle and the scanty remains of Roxburgh Castle, which to some extent still indicate the

Roxburgh Castle

characteristics of the Norman era—a dominant natural situation with massive embattled walls enclosing a large fortified area. Fourteenth-century constructions were of a different type, consequent upon the impoverished condition of the country following the War of Independence, and scarcely less, the fear that if great fortresses were erected, these might be captured and held as bases by the enemy. This was the period of the oblong form of keep known as the "peel" or "pele." The peel originally meant a palisaded or stockaded fortification of the "mote and bailey" type. The mote was a hillock of earth with steep sides rounded by a deep trench. The bailey was an attached enclosure at a lower level. On the hillock stood a wooden castle or *bretasch* with a palisade. By and by stone took the place of timber. The palisade itself might be reproduced by flanking stone walls if the site lent itself to that or not. Gradually, however, a single rectangular plan prevailed without any attendant outworks. To this there was often added, when more accommodation was required, a short wing built at one angle of the oblong keep, thereby converting it into a building resembling the letter L, from which this type of castle borrows its name. An Act of 1535 ordered every Borderer having land to the value of £100 to build a "barmkyn" and tower 60 feet square as a refuge for himself and his tenants. Those having smaller rentals were to construct "peels" for saving themselves and their goods. This form of building usually begins with a vaulted basement for storage, and for cattle when necessary. The first floor was the common hall with wide, ample fire-place. A second and a third floor were the domestic quarters. The uppermost floor, finished off by a

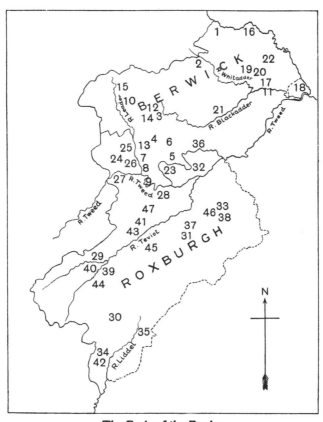

The Peels of the Border

THE PEEL TOWERS OF THE BORDER

1	Cockburnspath	13	Whitslaid	25	Glendearg	37	Timpendean
2	Cranshaws	14	Bassendean	26	Langshaw	38	Whitton
3	Evelaw	15	Carfrae	27	Darnick	39	Goldielands
4	Corsbie	16	Fast Castle	28	Littledean	40	Branxholme
5	Mellerstain	17	Edington	29	Harden	41	Barnhill
6	Greenknowe	18	Mordington	30	Hermitage	42	Mangerton
7	Rhymer's Tower	19	Bunkle	31	Ferniherst	43	Bedrule
8	Cowdenknowes	20	Billie	32	Roxburgh Castle	44	Crumhaugh
9	Bemersyde	21	Nisbet	33	Cessford	45	Fulton
10	Old Thirlestane	22	Houndwood	34	Hartsgarth	46	Corbet
11	Hutton	23	Smailholm	35	Lariston	47	Lanton
12	Wedderlie	24	Colmslie	36	Hume		

bartizan and roof of massive rough stone, was occupied by the garrison. An iron "yett" (gate) and narrow turnpike stair gave access to the whole. The walls were from seven to nine feet thick, and there was no attempt at ornamentation, the building being designed for strength and not for beauty.

Thus it is that the Border became studded over with these innumerable ruined and unruined peels without which much of its picturesqueness and its romance would have been impossible. Everywhere we meet with them— near a town or village, within fertile glens and remote "hopes," and mountain passes. Perched on some craggy knoll at whose base the burn meanders; sitting pleasantly on the peninsula between two uniting rivers; upreared on a mound or morrain at the valley-mouth; or posted on a projection of the hillside—wherever they stand, the main thing about them is the spirit of defiance they seem to proclaim to the world.

In Berwickshire, Cranshaws is a rectangular tower with rounded corners measuring 40 by 26 feet, and five storeys high, with a parapet supported on a single row of corbels. It is the only keep still in its original condition, standing alone and inhabited. Others, at Bemersyde, Cowdenknowes, Hutton, Wedderlie, Nisbet, Houndwood, though occupied, have been merged into domiciles of later date and otherwise considerably altered. The ruined keeps of the county number nineteen: Billie; Blanerne; Borthwick Castle, near Duns, a complete ruin; Bunkle; Carfrae; Cockburnspath—a likely enough prototype of the "Ravenswood" of the *Bride of Lammermoor*; an old structure overhanging the Leader at Cowdenknowes; Edington, in Chirnside

Cranshaws Tower

parish; Fast Castle, like an eagle's eyrie on a rocky bluff jutting into the sea for a thousand yards; the Rhymer's Tower at Earlston, the most literary of the keeps of Scotland; Greenknowe, home of Pringle the Covenanter; old Thirlestane on the Boonreigh Water; Whitslaid on the Leader; Edrington, in Mordington, a mere fragment incorporated in the modern farm-buildings; Evelaw and Bassendean, in Westruther parish; Corsbie, in Legerwood; Redbraes Castle at Polwarth; and a doubtful relic called Whiteside Tower, at Mellerstain. Greenknowe is the only complete keep of the L-plan now surviving in the shire. It is a domestic dwelling dating from 1581 and retaining almost none of the defensive features which characterised the earlier castles. The doorway is at the re-entering angle, giving access on the ground floor to the vaulted kitchen and to the wheel-stair formed in the L wing which communicates with a hall on the first floor, whence a projecting turret stair leads to the two upper floors. The wall head is finished with a plinth, the parapet being abolished, and the gables terminate with crow-steps and massive chimney-stacks. The iron "yett" at the entrance is in position, but the corbelled angle-turrets are little more than ornamental survivals of the earlier defensive features. The larger wing measures 24 feet by 15 within walls averaging 4 feet in thickness. The shorter wing is 15 feet 6 inches in width with a projection of 10 feet 4 inches eastwards.

Of the ruined peels of Roxburghshire, Allanhaugh, strongly posted on a scaur of the Allan Water, stood opposite Newmill-on-Teviot. Only a few fragments remain. Visible from many points, Goldielands is the most conspicuous landmark of upper Teviotdale. It was originally a watch-tower for Branxholme. Barnhills Castle, near the base of Minto Crags, Crumhaugh-on-Teviot, Lanton, Timpendean, Bedrule, Fulton, Cessford, Corbet on Kale, Whitton, Wallace's Tower at Roxburgh, the three towers

Littledean Tower

of the *Monastery*—Glendearg, Colmslie, Langshaw—Smailholm and Littledean, are best known. The last is unique in form. Its eastern side is the usual oblong and its western is crescent-shaped.

Smailholm is the most noted of these old defences. Despite its ruinous condition it is a most perfect relic of Border feudality. Steep, almost perpendicular crags hug three of its sides. A fourth overlooks a deep little loch, perhaps the lochan of *The Abbot.* The building is of the same dimensions as its neighbour at Bemersyde—24 feet 6 inches by 16 feet 6 inches within walls which are from

Smailholm Tower

7 to 9 feet thick. From base to balcony it is 60 feet, reached by a well-preserved spiral stair. There is no trace of ornamentation.. The iron "yett" is in position. The whole structure suggests immense strength. At Sandy-knowe farm-house, within a stone's throw, part of Scott's boyhood was spent, and the finest of his poetic word-pictures (Introduction to Third Canto of *Marmion*) was inspired by his early memories.

Darnick Tower is still inhabited. Built by a Heiton in 1425, it remains in the hands of a Heiton. Hertford burned the place in 1545, and it was rebuilt in 1569.

Of Fisher's Tower, close by, only three of the outer walls remain. The rest of the building is modern. Fernihirst, charmingly placed on Jed Water, was built in 1498. It has been much modernised and altered. Cessford Castle was one of the strongest fortresses of the Border. Surrey failed to capture it. One of its walls is 13 feet thick. The deep moat which environed the whole fabric is barely traceable.

Liddesdale was *par excellence* the Border peel-district. It was the country of Elliots and Armstrongs, of Nixons and Croziers—thieves all—

> Fierce as the wolf, they rushed to seize their prey,
> The day was all their night, the night their day.

Most of their ancient strengths have disappeared, though the sites are still pointed out. On the Liddel were Lariston, the Elliot fountain-head; Mangerton; Syde; Park; Copshaw; Westburnflat; Whithaugh; Clintwood; Hillhouse; Thorlieshope; Gorranberry, and Millsholm. On the Water of Hermitage stood Hartsgarth, Redheugh and Roan. Near the junction of Liddel and Hermitage were the castle of the "Lords of Lydal," and the important township of Castleton—not unlike the Roxburghs between Tweed and Teviot. Only the worn pedestal of Castleton Cross, and a deserted kirkyard have been left to tell the tale. Hermitage alone retains something of its old-world gloom, hidden in the heart of the hills, cursed by black and bloody memories. Built by the Comyns in the thirteenth century it passed to the Soulises, the Angus Douglases, to "Bell-the-Cat" himself, the Hepburn Bothwells, the "bold Buccleuch," and it is now under the care of His Majesty's Office of Works. Hermitage is in the form of a double tower. Its west front is 103 feet long, and the circumference of the whole about 600 feet. It is 60 feet high. The walls

Hermitage Castle

are 9 feet thick, and the top is surmounted by a projecting battlement. Under its new supervision, dilapidations both exterior and interior, may be expected to receive whatever retrieval is possible. It was at Hermitage that Sir Alexander Ramsay was starved to death in 1343. Here Queen Mary paid her flying visit to the Earl of Bothwell.

19. Architecture— (c) Domestic and Municipal.

When quieter days came with the Union of 1603 (though there was still much predatory warfare upon the Border), Domestic architecture began to develop. Defensive requirements ripened into humaner needs which deepened with the peaceful settlement of the country and engagement in agricultural pursuits that followed. The castle lost its military significance. From a garrison headquarters it grew into a home. Building became common. Provision was made for the amenities arising out of the new conditions. A taste for gardening was fostered. Ornamentation, necessarily overlooked amidst the old embroilments and forebodings, started on its triumphant way. Many of the peels were transformed into stately mansions. The slowness, however, of their evolution is apparent in the picturesque combination of Renaissance and native style, so characteristic of Scottish architecture during the seventeenth century.

Only the most important mansions can be mentioned.

In Berwickshire, one of the oldest is Thirlestane Castle, the seat of the Earl of Lauderdale. Originally erected by Chancellor Maitland on the site of the ancient Lauder Fort belonging to Edward I's reign, Thirlestane was really a creation of the Duke of Lauderdale, whose reconstruction of the Chancellor's house left it much in its modern condition. The arrangement consists of a central oblong block measuring 108 feet by 22, with a large rounded tower at each angle and six semicircular towers projecting at irregular intervals from the side walls, three on each side, the central pair containing wheel-stairs giving direct access from the basement to the floors above. The whole structure is unusual and arresting and has been likened to "a palace of the *Arabian Nights* in all the magnificence of its gorgeous

Thirlestane Castle, Lauder

display in art and classic adornment, enriched by the natural beauties of a unique situation."

Cowdenknowes, near Earlston, possesses what is perhaps the loveliest location on Leaderside. Ancient and modern features blend in happy harmony in the modern mansion. The original house was of the castellated type, with flanking fortalices connected by curtain-walls enclosing the usual courtyard. In the Hertford devastations Cowdenknowes was destroyed. On a reconstruction, the old keep, which bears the date 1554, became a separate building. Through this a handsome doorway admits to the re-united fabric— an inner hall occupying the former open space between the two structures.

Mellerstain, within Earlston parish, is the princely seat of the Earl of Haddington. Built in 1725 and enlarged at

Cowdenknowes, Earlston

a later date from plans by the elder Adam, in a style long popular in Scotland—a central block with two wings—it has undergone considerable modification, and is now one of the most attractive residences on the Border.

Spottiswood, occupying a commanding site in Westruther parish, was re-erected in 1830 in the Elizabethan style. Over the window of an entrance-lodge is a pediment from

the Glasgow manse of Archbishop Spottiswood with date
1596 and an inscription: *Mihi vivere Christus et mori
lucrum.*

Mertoun House, built in 1702 from designs by Sir
William Bruce, was reconstructed in 1913. It is a plain
classic edifice on a picturesque nook by the Tweed. For
long the seat of the Scotts of Harden—of Lord Polwarth,

Mertoun House

chief of the clan—it is now the property of the Earl of
Ellesmere. Scott was a frequent guest here. For its laird
he wrote the *Eve of St John*, and "Christmas on Tweedside" in *Marmion* is an exquisite pen-picture of the hos-
pitable customs and ceremonies of a by-past age. In the
flower-garden is one of the oldest dovecots in Britain, of
date 1576.

Bemersyde is probably the most ancient dwelling in the

Border, and one of the few in Scotland still held by a descendant of its original owner. Its early keep, of the same type and date as Cranshaws and Smailholm, is now the central block of a modern mansion. Deep interest pertains to Bemersyde from its association with Thomas the Rhymer's prophecy (p. 182), and from its significant fulfilment in modern times when a grateful nation gifted to

Bemersyde

Field-Marshal Earl Haig the abode of his "long descended line."

Duns Castle is a stately and imposing baronial pile, and has an interesting history behind it. In 1314, Robert the Bruce gave Duns to his nephew Randolph, Earl of Moray. About 1316 Randolph built the original castle of which the only existing portion is a tower at the east end incorporated in the present edifice. In 1346 Duns Castle was

Duns Castle

in the possession of Randolph's daughter, the redoubtable Black Agnes of Dunbar.

Nisbet House is a good example of the seventeenth century mansion, its dominant features being lofty circular towers formed at the southern angles and projecting stair-case wings to the north, with shot-holes, which are purely ornamental.

Hutton Castle is another Berwickshire mansion completely modernised. It retains the original keep, which projects at the south-east end of the house. Measuring 30 feet by 24 feet, it is three storeys high, with a circular tower (heightened) containing the staircase.

Ayton Castle is a magnificent baronial edifice of red stone built in 1856 on the site of a predecessor destroyed by fire in 1834.

Ninewells is an elegant Tudor mansion in Chirnside parish, built in 1841, successor to an older house belonging to the Humes. The Hirsel, a seat of the Earl of Home, is a spacious sandstone mansion in Coldstream parish. Portions appear to date from the seventeenth century. Langton House, erected in 1862, on the site of an older edifice, was a stately Elizabethan structure surrounded by extensive grounds with a noble entrance gateway of 1877. Marchmont House was built by the last Earl of Marchmont in 1754 to replace the original Redbraes Castle. Designed by Robert Adam, it is of semi-Palladian style in white sandstone, and has an imposing appearance. Considerable adjustments have been effected within a recent period, and the place is no longer a Hume domain. Swinton House is an elegant modern mansion, successor to one of great antiquity. It is still owned by a Swinton. Wedderburn Castle is in the Greek style. Kimmerghame is a handsome mansion in Scottish Baronial style erected in 1851. Wedderlie House, in Lammermoor, is a typical example of the evolution of the ancient keep. Built about the beginning of the seventeenth century, it consisted of two wings, to the east of which a dwelling-house has been added. The form is oblong with a frontage measurement of 76 feet 6 inches, and 16 feet 6 inches in width within walls.

Other mansions are: Anton's Hill, Bassendean, Belchester, Blackadder, Blanerne, Broomhouse, Caldra, Carolside, Dryburgh, Eccles, Edrington, Edrom, Gledswood, Kames, Kelloe, Ladykirk, Lees, Lennel, Longformacus, Manderston, Milne-Graden, Mordington, Nenthorn, Newton-Don, Paxton, Peelwalls, Purveshall, Renton, Rowchester and Whitchester.

Of Roxburghshire mansions Abbotsford is the most celebrated. Scott, suffering from "yerd-hunger," purchased Cartleyhole in 1811—a dilapidated farm-house, with 100 acres of land. It was a most uninviting spot whereon to rear a stately dwelling. How he succeeded in creating out of it his marvellous aggregation of stone and lime, his broad and fair territory of field and farm, is known to everybody. A description would be out of place here. The edifice was completed by 1825. About 1853, Mr Hope-Scott, husband of Lockhart's daughter, spent huge sums on additions and improvements. This extension, in light freestone, is easily recognisable in contrast to the darker hue of Sir Walter's house, which was built of native blue whin. Abbotsford was reared on no set plan, but with the desire to reproduce some of those features of ancient Scottish architecture which Scott most venerated. It is at once

Abbotsford

a monument of the high historical imagination from which sprang his more enduring memorial, and of the over-zeal which may be lavished with very disastrous results, on the mere "pomp and circumstance of time"—the all-absorbing passion

> To call this wooded patch of earth his own,
> To rear a pile of ill-assorted stone
> And play the grand old feudal lord again.

The ducal palace of Floors is without a rival in the south of Scotland. Sir Walter Scott wrote that "the modern mansion of Fleurs, with its terraces, its woods, and its extensive lawn, form altogether a kingdom for Oberon or Titania to dwell in, or any spirits who, before their time, might love scenery, of which the majesty and even the beauty impress the mind with a sense of awe mingled with pleasure." Modern Floors is the result of a vast scheme of reconstruction and ornamentation in 1839 upon the old house built in 1718 for the first Duke of Roxburghe by Sir John Vanbrugh, dramatist and architect.

Springwood Park is the "idyllic" seat of Sir George Douglas, Bart., Border poet and historian. Built in the Palladian style, the original structure was completed in 1756, and enlarged in 1852. An imposing gateway, designed by Gillespie Graham, gives entrance to a park of much natural beauty.

Hendersyde in Ednam parish, built in the style of an Italian palazzo, is the seat of Sir Richard J. Waldie-Griffith, Bart. It contains a valuable Library. An extensive park abounds in "a brotherhood of venerable trees."

Monteviot, a seat of the Marquess of Lothian, sits romantically at the base of Penielheugh Hill (774 feet)

Floors Castle

with its far-seen landmark—a cylindrical pillar 156 feet high, erected as a memorial of Waterloo.

Cavers is a noteworthy Teviotdale house, but so much altered that its original form is not easy to expiscate. There are the remains of at least five different periods of building —traces of the early Baliol house; the tower of the Wardens of the Middle Marches; the mansion of the Sheriffs of Teviotdale at the time of the Union; the Italianised composition of 1763; and later extensions. In the twelfth and thirteenth centuries Cavers was owned by the Norman Baliols. Hugh de Baliol was one of those who witnessed the Peace Treaty signed at Cavers in 1237. About 1360 the place was inherited by the first Earl of Douglas. The gauntlets taken at Otterburn from Percy, with the Douglas pennon recovered from the battle, are still in the custody of his successor.

Harden exhibits no architectural features of note. It is a severely plain edifice, but no Border house holds such memories, or has played a greater part in the storm and stress of Border warfare. Here lived the renowned Wat o' Harden, "king of the reivers," whose deeds, dark and splendid, Scott tells us he made to ring in many a ditty. Sir Walter was a direct descendant of "auld Wat" and of his no less famous spouse, Mary Scott, the "Flower of Yarrow."

Minto House, the residence of the Earl of Minto, was built by the first Earl who did not live to see its completion in 1814. It is a handsome four-storeyed edifice at the head of Minto Glen and fronted by an artificial lake. Jean Elliot, writer of the Scottish version of the *Flowers of the Forest*, was born at Minto in 1727.

Other Roxburghshire mansions are: Allerly, Ancrum House, Benrig, Drygrange, Ednam, Edgerston, Gattonside, Greenriver, Greycrook, Hallrule, Eildon Hall, Hartrigge, Huntlyburn, Kippilaw, Kirklands (Ancrum), Lariston, Lessuden, Lowood, Maxpoffle, Maxton, Pavilion,

Harden

Ravenswood, Riddell, St Boswells Bank, Stirches, Stobs, Threepwood, Wauchope, Weens, Wells, Wolfelee, and Wooplaw.

In Berwickshire no town is sufficiently large and wealthy to warrant the erection of public buildings on any extensive scale. Many of the houses in Duns are old and quaint, but

Duns

Old County Buildings, Greenlaw

modern architecture is well represented. In its spacious Square is the Town Hall, of Gothic design, small, sufficient. Greenlaw has a somewhat ornate edifice in the Grecian style long used for county purposes. Intersecting

Queen Mary's House, Jedburgh

Lauder Street is the old Tolbooth of the burgh, a Dutch-looking structure whose upper storey does duty as Council Chamber, the under storey as a jail. Coldstream Town Hall was gifted by the Earl of Home. A statued pillar 70 feet high in memory of Charles Marjoribanks, M.P., is a conspicuous object. Eyemouth Town Hall is a handsome Romanesque edifice.

Roxburghshire possesses more buildings of a public character. In Hawick the Municipal Buildings were erected at a cost of £13,000 in the Scottish baronial style. The Science and Art Institute in Italian Renaissance, is a

Kelso Bridge

memorial to the Duke of Buccleuch. In Jedburgh Queen Mary's House, three-storeyed, red-roofed, where Mary lay deadly sick after her Hermitage adventure, is a rare link with the Past. The County Buildings and a massive memorial of the Great War are prominent structures. Kelso Bridge, a noble span of five elliptical arches, each of 72 feet, was built by Rennie in 1803, and became his model for Waterloo Bridge, London.

20. Communications—Roads and Railways.

The Communications of a county are the lines along which traffic is transported from point to point, without or within, and over which passengers are conveyed: generally the easiest and shortest route is followed. This depends on the physical conditions, on the relative situation of places, and on the character of the soil so far as constructive operations are concerned. Land communications by road and rail are those which obtain in the Border Counties, where there are no canals or lakes usable as waterways. The most obvious routes are the river-valleys, and through these the main highways of the district have been carved out.

The Romans were the first scientific pathfinders in our island. A tortuous track on the hillside and through the forest was all that the Border tribesmen knew. On the other hand, the Romans were never daunted by the natural barriers. They swept their broad paved roads along straight routes, over hill and dale, through scrub and marsh. Many of these roads still remain, and some are incorporated in the road surfaces now in use. The principal routes into Scotland from the south were Watling Street, an unquestionably Roman construction for most of its course upon English soil, though not so decided after it has crossed the Cheviots, where its character changes into a more or less winding ridgeway. It can be traced to Cappuck Camp on Oxnam Water, and onward to Jedfoot Bridge and Ancrum. From Newstead it is supposed to have followed

the Tweed to Melrose and Darnick, and to have struck
north by Blainslie and Lauder to Soutra Hill and Midlothian. "Derestreet" is only another (and, Haverfield says,
the correct) name for Watling Street. In various charters
its position is indicated in such a way as seems to leave
no doubt of its identification with the Roman highway.
A Melrose Abbey charter has a reference to the King's
Way (Regis Via) "which goes from Annandale towards
Roxburgh." This ancient thoroughfare ran up the valley
of the Liddel, crossed the summit dividing Liddesdale from
Jedwater (being identical for part of its course with the
road known as the Wheel Causeway), thence it came down
into Rulewater for a few miles and striking off towards
Swinnie and Swinnie Moor, entered the Castlegate of Jedburgh by a road now represented by the Loaning. From
Jedburgh it proceeded to Roxburgh and up Tweed to
Lauderdale, being identical with Derestreet from about
St Boswells in its northward track. It is practically certain
that Watling Street, Derestreet, the King's Way, and what
was known as "Malcolm's Road" were parts of one and
the same road leading from the Border to the North.

Several "herring" roads crossed the Lammermoors into
Berwickshire. The Girthgate is the name of an ancient
roadway passing through Channelkirk towards Soutra. It
is said to have gone down the Elwyn to Melrose, but no
doubt its true destination was the "girth" or sanctuary at
Soutra Hospital.

The present roads of the counties are well constructed,
well kept, together with the bridges over which they pass.
Large sums are spent, and considerable grants are made by
Government. Immense quantities of metal for "bottom-

ing," thousands of tons of sand, and hundreds of barrels of tar-bitumen, are absorbed annually in the reparation of every main highway.

The chief routes traversing Berwickshire are the Great North Road from London to Edinburgh, which enters the county at Lamberton, and passes Burnmouth, Ayton, Houndwood, Grantshouse and Cockburnspath. Its distance is 19 miles. The north road from Berwick to Kelso passes through Swinton, Leitholm, and Eccles. The road from Coldstream to Kelso runs parallel with the Tweed. Roads from Coldstream to Haddington go through Duns and Longformacus; from Coldstream to Lauder through Greenlaw; from Eyemouth to Lauder through Ayton, Chirnside, Duns and Westruther. The east road from Kelso to Edinburgh passes Earlston and Lauder, leaving the county at Soutra.

The chief Roxburghshire roads are those from Hawick to Jedburgh and Kelso by Crailing or the Dunian; Hawick to Newcastle *via* the Carter; Hawick into Liddesdale by Limekilnedge; Hawick to Teviothead; Jedburgh into Liddesdale by Note (or Knot) o' the Gate; Hawick to Selkirk; Kelso to Earlston by Smailholm; Kelso to Earlston by Maxton, St Boswells and Leaderfoot; Earlston to Galashiels through Gattonside; and Earlston to Melrose through Newstead.

Prior to 1764 there were only two bridges over the Tweed—at Melrose and Kelso, and two over the Teviot—at Hawick and Ancrum. Now every stream is bridged. Ferryboats cross the Tweed at several points. Many smaller towns have been linked up by a system of motor-conveyances.

The only railway system in Berwickshire is the London and North Eastern, formerly the North British. This line, running along the coast from Edinburgh, enters the county near Cockburnspath station and continues in a south-easterly direction to Berwick, with stations at Grantshouse, Reston, Ayton and Burnmouth. From the latter place, a short branch, opened in 1891, leads to Eyemouth. From the main line at Reston, a line leads across the county to Earlston and St Boswells, with stations at Chirnside, Edrom, Duns, Marchmont, Greenlaw and Gordon. From Duns to Earlston, this, opened in 1863, was originally known as the Berwickshire Railway. A further extension to St Boswells was made in 1865 and in 1876 the whole was merged in the North British Railway system. A light railway branching off from Fountainhall in Midlothian on the main Waverley Route from Edinburgh, and continuing to Oxton and Lauder has opened up since 1901 with signal success the wide agricultural district of upper Lauderdale.

The Waverley route of the L. and N.E. Railway from Edinburgh to Carlisle runs for about 50 miles through Roxburghshire, that county being entered shortly after leaving Galashiels. There are stations at Melrose, St Boswells, Belses, Hassendean, Hawick, Stobs, Shankend, Riccarton, Steele Road, Newcastleton, and Kershope Foot. St Boswells is the junction for the Berwickshire branch. From St Boswells another line proceeds by the Tweed to Kelso with stations at Maxton, Rutherford, Roxburgh, where a branch is sent off southwards to Jedburgh with stations at Kirkbank, Nisbet and Jedfoot Bridge. From Kelso the line is continued by the English North-Eastern to Berwick, with stations at Sprouston, Carham and Sunilaws. Cornhill is the station for Coldstream. From Riccarton, on the main line, a branch goes eastwards, passing the station of Saughtree before leaving the county for Hexham and Newcastle.

21. Administration and Divisions.

The office of Sheriff in Scotland is first mentioned in the beginning of the twelfth century. The Sheriff was the minister of the Crown, responsible for all civil and criminal cases. Within his sheriffdom his authority was unlimited. He had no jurisdiction beyond his own district, in which he was bound to take all means to have the law proclaimed that no one might pretend ignorance. At an early date the office became hereditary and continued till soon after the second Jacobite Rebellion. The office was generally held by a member of the Scottish nobility, and it was not only hereditary but could pass by sale with the estate to which it was attached. An indirect result of this was that the hereditary Sheriff, frequently incapable of discharging the duties, had to find a Depute, which explains the origin of the name, Sheriff-Depute.

The counties are governed by a Lord-Lieutenant, who may be a commoner. Under him is a Vice-Lieutenant with (1925) thirteen Deputy-Lieutenants for Berwickshire and sixteen for Roxburghshire. Judicial procedure is in the hands of the Sheriff-Principal, who is non-resident, and holds a triple appointment as Sheriff of Berwick, Roxburgh, and Selkirk shires. There is, however, a resident Sheriff-

Substitute for each of these shires as well as a number of Honorary (unpaid) Sheriff-Substitutes. Criminal prosecution is the business of a Procurator-Fiscal who has his Deputy. Sheriff-Courts are held at Duns, Jedburgh, and Hawick. There are (1925) over 300 Justices of the Peace for Berwickshire and Roxburghshire, including several women. Justice of the Peace Courts and Small Debt Courts are held at Ayton, Coldstream, Duns, Lauder, Jedburgh, Kelso, Hawick, and Melrose. Licensing and Licensing Appeal Courts meet at Duns and Jedburgh. The Burgh of Hawick has its own Appeal Court. Each county has its own Chief-Constable, and an Inspector of Weights and Measures. Inland Revenue matters are controlled from Galashiels.

The chief administrative body is the County Council. Under the Local Government (Scotland) Act (1889), Berwickshire is divided into 33 Electoral divisions, which return 33 members to the Council, whose chairman and vice-chairman are designated Convener and Vice-Convener. There are three Districts in the County—the East, with 11 electoral divisions, the Middle, with 14 electoral divisions, and the West, with 8 electoral divisions.

Roxburghshire has 28 Electoral Divisions, returning 31 members to its County Council, and there are five Districts in the area—Kelso with 9, Jedburgh with 7, Melrose with 5, Hawick with 5, and Liddesdale with 2 electoral divisions. Each of these eight Districts has a District Committee consisting of the County Councillors for the electoral divisions, and of parish councillors chosen by each Parish Council of the district. There are thus two representatives on the District Committee—one elected by the ratepayers, the other appointed by the parish council. Administration of the Public Health Acts is the chief business of the District Committee. It cannot levy assessments and has no authority over finance. The powers of the Council are the maintenance of roads and bridges, the administration of the Contagious Diseases (Animals) Act, the appointment of medical officers of health and sanitary inspectors, an agricultural analyst under the Fertiliser and Feeding Stuffs Act, and a public analyst under the Sale of Food and Drugs Act. Pollution of rivers and protection of wild birds also come under their purview.

Burghs are governed by Town Councils, consisting of a Provost, Magistrates, and Councillors who retain their seats for three years. In Berwickshire, the burghs of Coldstream, Duns and Lauder have each nine members. Eyemouth has four members. In Roxburghshire, Hawick and Jedburgh have fifteen members, Kelso twelve and Melrose eight. Lauder and Jedburgh are the only Royal Burghs in the shires. The others are police burghs.

The remaining parishes in Berwickshire are Abbey St Bathans, Ayton, Bunkle, Channelkirk, Chirnside, Cockburnspath, Coldingham, Cranshaws, Earlston, Eccles, Edrom, Fogo, Foulden, Gordon, Greenlaw, Hume, Hutton, Ladykirk, Langton, Legerwood, Longformacus, Mertoun, Mordington, Nenthorn, Polwarth, Swinton, Westruther, Whitsome: and in Roxburghshire—Ancrum, Bedrule, Bowden, Castleton, Cavers, Crailing, Eckford, Edgerston, Ednam, Hobkirk, Hownam, Kirkton, Lilliesleaf, Linton, Makerstoun, Maxton, Minto, Morebattle, Oxnam, Roberton, Roxburgh, St Boswells, Smailholm, Southdean, Sprouston, Stichill, Teviothead, and Yetholm.

Prior to 1872 education was in the hands of the clergy of the Church of Scotland and the heritors (landed proprietors). An Education Act of that year introduced an elective School Board system. This in turn passed with the Education Act of 1918, which instituted another elective body, the County Education Authority. The parish is no longer the administrative unit of education, but members are returned by the County electoral divisions. Berwickshire Education Authority consists of 14 members; that of Roxburghshire has 24 members. The former meets at Duns, the latter at Newtown St Boswells. Education is free and compulsory for all children between five and fourteen, with conditional exemption in agreed-upon circumstances. Provision is made for the institution of nursery schools for children between two and five, and a generous system of continuation classes for pupils above fifteen has been made possible. Medical supervision is amongst other benefits. With adequate financial aid, wherever expedient, opportunity is now given to every child to pass from the rural to High School and from High School to Training College or University. Scotland has been celebrated for its educational system from the days of John Knox, who aimed at a school in every parish throughout the land. A greater dream has been realised to-day.

Berwickshire unites with East Lothian in returning one representative to Parliament. In 1924 the electorate numbered 34,017. Roxburghshire and Selkirkshire (including the Hawick Burghs), with a parliamentary constituency of 34,529, return one member.

Ecclesiastical affairs are managed by 32 Kirk Sessions belonging to the Church of Scotland, in Berwickshire, and by 25 belonging to the United Free Church. Hound-wood is the only *quoad sacra* parish in the county. The Church of Scotland Presbyteries are those of Duns, Chirnside and Earlston. In Roxburghshire there are 30 Kirk Sessions belonging to the Church of Scotland, and 27 belonging to the United Free Church. The *quoad sacra* parishes are six—Edgerston, Hawick (St John's, St Mary's, St Margaret's), Kelso (North) and Saughtree. The Church of Scotland Presbyteries are those of Kelso and Jedburgh. Nenthorn, though in Berwickshire, is in the Presbytery of Kelso; Smailholm, though in Roxburghshire, is in the Presbytery of Earlston. Stow, a wholly Midlothian parish, is also in that Presbytery. Bowden, Lilliesleaf, Maxton, Melrose, Robèrton, St Boswells, all in Roxburghshire, are in the Presbytery of Selkirk. Cockburnspath, in Berwickshire, and within the Synod of Lothian and Tweeddale, is in the Presbytery of Dunbar. With this exception, all other parishes mentioned are within the bounds of the Synod of Merse and Teviotdale.

The Episcopal Church in Scotland has congregations at Chirnside, Coldstream, Duns, Eyemouth, Hawick, Jedburgh, Kelso, and Melrose. The Roman Catholic Church conducts services at Duns, Hawick, Jedburgh and Kelso. The Congregationalists have five places of worship, and the Baptists three. The Free Church of Scotland has no representation within the Border area.

22. The Roll of Honour.

A conspectus of Border men and women whose names are
deemed worthy of being inscribed on the bede-roll of fame
would occupy a fair-sized volume. In proportion to their
size and population Berwickshire and Roxburghshire have
produced more notable natives and have had more eminent
persons connected with them than any other part of Scotland.

First of all, there are the great territorial families of the
counties—in Berwickshire, the Homes, Maitlands, Spottiswoods, Cockburns, Erskines, Swintons, Baillies, and Gordons: in Roxburghshire, the Scotts, Kers, and Elliots.

The Homes (pronounced Hume) sprang from the
doughty Gospatricks, Earls of Northumberland, one of
whom with other English nobles fled to Scotland after the
Conquest of 1066. They carried with them Edgar Athe-
ling, heir of the Saxon line, and his two sisters, Margaret
and Christian. Malcolm Canmore married Margaret,
bestowing on her protector Gospatrick the Earldom of
Dunbar and broad lands in the Merse and Lauderdale.
Patrick, son of the third earl, married his cousin Ada. By
her he obtained the territory of Home which thus became
the family designation. A later Home acquired Dunglass,
whence their second title of Lord Dunglass. Sir Alexander
Home accompanied the Earl of Douglas ("Tineman") to
France and fell at Verneuil in 1424. The second Lord
Home ruled Scotland as Lord High Chancellor in James
IV's time. The third commanded at Flodden, escaping its

carnage to suffer (unjustly) a traitor's death in 1516. The fifth turned the tide of battle at Langside against Queen Mary. The sixth received the Earldom of Home. Wedderburn was the most powerful sept—progenitor of nearly all the other Berwickshire Homes. The "seven spears of Wedderburn" are celebrated in song. David of Godscroft (1560–1630) was chronicler of the clan. Sir Everard (1756–1832), eminent surgeon, and Robert (1751–1834), portrait-painter, were of Greenlaw Castle stock. John Home (1722–1808), author of *Douglas,* was descended from the Cowdenknowes branch. Of the Polwarth Humes Sir Patrick (1556–1609) was antagonist in the *Flyting* with Alexander Montgomery. Alexander Hume (1560–1609) wrote *Hymns and Sacred Songs,* and a finely descriptive poem, *The Day Estivall.* He was minister of Logie, in Stirlingshire. The fortunes of the Polwarth line culminated in Sir Patrick (1641–1724), statesman and Covenanter. How he lay hid in a vault of Polwarth Kirk where his daughter Grisell (1665–1746)—a mere slip of a girl—attended to his wants is one of the classic tales of the Merse. Finding his way to Holland, he returned with William of Orange. His estates were restored and he was created Lord Polwarth, becoming Lord High Chancellor and Earl of Marchmont in 1697. Hugh, last Earl (1708–1794), was one of the wits of Queen Anne's time, the intimate and executor of Alexander Pope the poet.

Sir Richard Maitland, Lord Lethington (1496–1586), was a collector of Early Scottish poetry and himself a versifier. The Maitland Club was called after him. His son William (1528–1573) was the famous Secretary of Queen Mary's reign. Another son, John (1545–1595), was Lord High Chancellor and first Baron Thirlestane. He composed a satire *Against Sklanderous Tongues*. His grandson was that scourge of the Covenanters, the notorious Duke (1616–1682), only bearer of that title. The fourth

Lady John Scott [*A. E. Chalon, R.A.*

Earl (1653–1695) translated Virgil. The eighth earl (1759–1839) wrote on political economy. The tenth and eleventh earls were eminent naval officers—Admirals.

Several of the Spottiswoods filled important positions.

Superintendent John (1510–1585) officiated at the coronation of James VI at Stirling in 1567. James (1567–1645) held the Irish See of Clogher. Archbishop John (1565–1639), the historian, crowned Charles I at Holyrood in 1632. Robert (1596–1646) shone as a lawyer. He was executed for complicity with Montrose. Alicia Anne (1810–1900) wrote the ever-popular *Annie Laurie*. The Baillies derive their name from Bailleul in Normandy. The father of Robert Baillie, "the Scottish Sydney," purchased Mellerstain in 1643. Baillie, who was a kinsman of John Knox, perished on the scaffold in 1684 for alleged participation in the Rye House Plot. Whilst a prisoner at Edinburgh in 1677 his friend Hume of Polwarth desired to communicate with him and employed his daughter Grisell to convey a letter to the Tolbooth. In the performance of this task she had to consult with the prisoner's son George, who fell in love with her and married her 15 years later.

The Cockburns of Langton had a grant of that barony in 1595. Admiral Sir George (1772–1853) conveyed Napoleon to St Helena. Dr William (1669–1739) was medical adviser to Dean Swift. Sir Alexander (1802–1880), Lord Chief Justice of England, was a scion of this family. The Swintons have given many hostages to fortune. The first Swinton aided Malcolm Canmore in wresting his kingdom from Macbeth, and this incident no doubt accounts for the gift of the lands they still hold. They were a race of soldiers and statesmen. Sir John had a leading part in the capture of Hotspur at Otterburn. Under the more euphonious "Sir Alan," he is the hero of Scott's *Halidon Hill*. Another Sir John fought in France at Beaugé, where

Lady Grisell Baillie
From a painting at Mellerstain

he slew Thomas, Duke of Clarence, brother of Henry V. George (1750–1854) was President of the India Board of Trade. Swinton's Islands in the Mergui Archipelago of Lower Burma were called after him. Archibald Campbell (1812–1890) was Professor of Civil Law at Edinburgh. He wrote an illuminating history of his family. James Rannie (1816–1888) was the most fashionable portrait-painter of his time.

Other county families whose members rendered yeoman service to the state in peace and war, were the Haigs of Bemersyde; the Blackadders; the Edgars of Wedderlie; the Erskines of Shielfield, of whom were Ebenezer and Ralph, founders of the Secession Church; the Marjori-bankses, of whom was Dudley Coutts (1820–1894), banker and first Lord Tweedmouth; and the Nisbets, of whom was Alexander the Heraldist.

The "gay Gordons" had a Berwickshire origin. Gordon, their ancient patrimony, has long ceased association with the family, though the Duke of Richmond and Gordon is still "superior." At an early period they migrated from the Border—one set going into Galloway, whither they carried the name of Earlston; another set transferred the place-name of Huntly to Strathbogie, Aberdeenshire, where they blossomed as Earls, Marquesses, and Dukes.

The Scotts appear as Roxburghshire lairds in 1420, when Robert Scott of Buccleuch acquired the lands of Branxholme on Teviot. It is still held by a Buccleuch. Sir Walter, known as "Wicked Wat" defeated the English at Ancrum Moor in 1544. He was murdered by the Kers in the High Street of Edinburgh in 1552. Another Sir Walter, Warden of the Western March, was the most powerful baron of his time. One of his many exploits is commemorated in *Kinmont Willie*. Anne Scott, Countess of Buccleuch, the greatest heiress of her time, married the Duke of Monmouth, Charles II's son by Lucy Walters.

Lord Heathfield   [*Sir Joshua Reynolds*

She was in her twelfth year. The fifth Duke (1806–1884) was a leader in every enterprise for the betterment of the people. His statue stands under the shadow of St Giles' at Edinburgh. "Auld Wat" was the most famous of the Harden stock. His marauding deeds are household traditions. His wife was Mary Scott of Dryhope, the "Flower of Yarrow." The Raeburn Scotts—Sir Walter's branch—belong to the Harden sept.

The Kers, or Kerrs (pronounced and often spelled Carre or Karr) founded the families of Cessford and Ferniherst. John Ker, third Duke of Roxburghe (1740–1804) was the celebrated bibliophile. The sale of his Library lasted 45 days and at its close the Roxburghe Club was inaugurated. Robert Ker, first Earl of Ancram (1578–1654) was a man of cultivated tastes and a poet. Schomberg Henry Kerr (1833–1900), ninth Marquess of Lothian, was Secretary for Scotland.

The Elliots and Elliotts, "lions of Liddesdale," originally of Lariston and Redheugh, branched out into the lines of Stobs and Minto, of Wells and Wolfelee. George Augustus (1717–1790) was of the Stobs line. He is spoken of as the "Wellington of the Borders" from his defence of Gibraltar in the great siege which lasted from 1777 to 1784. He was created Lord Heathfield. His portrait by Sir Joshua Reynolds is in the National Gallery. The first Earl of Minto (1751–1814) was Governor-General of India. Jean Elliot (1727–1805) was authoress of the *Flowers of the Forest*, perhaps the finest specimen of Scottish elegiac verse. The fourth Earl (1845–1914), when Lord Melgund, commanded in the Riel Rebellion in Canada. He became in turn Governor-General of Canada and Viceroy of India. Sir Walter Elliot of Wolfelee (1803–1887) was a distinguished diplomatist.

To our counties the Court of Session has been indebted for many of its Judges. A few names may be mentioned—their legal titles: Lords Cessnock, Crossrig, Dirleton,

Edgefield, Harcarse, Hermiston, Jerviswood, Kames, Low, Mersington, Minto, Newabbey, Newhall, Nisbet, Ravelrig, Renton, Reston, Swinton, Tofts. David Hume of Ninewells was Principal Clerk of Session. John Hay

Jean Elliot
From a miniature at Minto

Athole Macdonald (Lord Kingsburgh), a Lord Justice Clerk, was of Ninewells stock.

Many eminent churchmen have been connected with our counties. Foremost stands the name of Cuthbert (637–687), great Border saint and evangelist of Tweedside.

Constant tradition makes him a native of Channelkirk. At all events, when he is first heard of, it is as a shepherd-lad on Leaderside. In 651, while tending his flocks by night be believed he saw the heavens opened and an angelic host carrying to glory the soul of the holy Bishop Aidan.

Statuette of St Cuthbert
in a niche at Melrose Abbey

As a result Cuthbert became a monk at Old Melrose, to close his career as Aidan's successor.

Others who may be mentioned are: Archbishops Blackadder, Burnett (Lauder), Cairncross (Colmslie), Forman (Hutton); Bishops David de Bernham (Nenthorn), Shoreswood (Bedshiel), Stewart (Minto) and Turnbull (Minto), founder of Glasgow University.

Many eminent ministers of religion are of later date. William Fowler (1560–1612), minister of Hawick, uncle of Drummond of Hawthornden, was Secretary to James VI's Queen and wrote *The Tarantula of Love* and *The*

John Cairns, D.D., LL.D.

Triumphs of Petrarch. Thomas Boston (1676–1732), a native of Duns, wrote *The Fourfold State.* Dr David Bogue (1750–1825), founder of the London Missionary Society, belonged to Coldingham. Dr John Cairns (1818–1892),

Principal of the United Presbyterian College, was born in a shepherd's cottage at Ayton Hill. He might have been Principal of Edinburgh University but told no one of the offer until that fact was revealed by his papers after his death. Dr Alexander Waugh of London (1754–1827) belonged to Gordon. James Murray (1732–1782), author of the curious *Sermons to Asses*, was born at Fans, Earlston. Dr Alexander Hislop (1845–1908) hailed from Earlston. Dr Thomas McCrie (1772–1835), biographer of Knox and Melville, and Dr John Duns (1820–1909), scientist and biographer, belonged to Duns; Dr Patrick Fairbairn (1805–1874) and Dr George Smeaton (1814–1889) exegetical scholars, to Greenlaw; Dr Adam Thomson (1779–1861), who was the means of breaking down the Bible monopoly, to Coldstream. Dr Andrew Martin Fairbairn (1838–1912) was of Lauderdale extraction. Among Roxburghshire divines, the famous Samuel Rutherford (1600–1661) was a native of Nisbet. David Calderwood (1575–1650), ecclesiastical historian, was minister of Crailing. John Livingston (1603–1672), Covenanter, was minister of Ancrum. The grandfather of Principal Robert Herbert Story was schoolmaster at Yetholm, and had a gypsy strain in his blood.

A number of men who attained distinction as pioneer missionaries should be mentioned: George Archibald Lundie laboured in Samoa; Archibald William Murray in Polynesia; Dr John Hume Young in Amoy. Mellerstain gave George Ainslie to a remarkable career amongst the North American Indians. James Gray (1770–1830) was rector of Dumfries Academy. Becoming imbued with the missionary spirit he spent his later years in India. He was a poet of some note. Stephen Hislop (1817–1863), who, like Gray, belonged to Duns, was one of the greatest of Indian missionaries. Hislop College at Nagpur commemorates his noble services. Dr John Wilson (1804–1875) was a native of Lauder. He went to India in 1828, became a great Oriental scholar and a Fellow of the Royal Society. His influence radiated from Bombay over the whole of India. The Wilson College was founded in his memory.

To Berwickshire and Roxburghshire belonged many of the sufferers and martyrs of Covenanting days. Among these were Sir Patrick Hume and Robert Baillie, already mentioned; Alexander Hume of Kennetsidehead, whose execution was the most cruel and unprovoked of the judicial murders which led to the Revolution of 1688; John Veitch, minister of Westruther; James Guthrie, minister of Lauder; Alexander Shields, author of the *Hind Let Loose*, and his brother Michael, compiler of *Faithful Contendings Displayed*, natives of Haughhead, Earlston; Walter Pringle of Greenknowe and Henry Hall of Haughhead, Eckford. James Kirkton, church historian, was minister of Mertoun. Robert Calder, minister of Nenthorn, had unenviable notoriety as author of the scurrilous Episcopalian tract, *Scotch Presbyterian Eloquence Display'd*.

Not a few leading medical men belong to both counties. From Bunkle came Dr John Brown (1735–1788), founder of the Brunonian system of medicine. Dr George Johnston (1797–1855), founder of the Berwickshire Naturalists' Club, and of the Ray Society, was born at Simprin. Dr Robert Dundas Thomson (1810–1864), noted physiologist and sanitation authority, was born at Eccles. Sir

John Pringle (1707–1782) of Stichill, sometime Professor of Moral Philosophy at Edinburgh, served as a doctor with the British Forces on the Continent. He attended the Duke of Cumberland during the "Forty-Five," and afterwards settled as a successful physician in London. His monument is in Westminster Abbey. Dr John Armstrong (1709–1779) is remembered more as a poet than as a man of physic. His *Art of Preserving Health* is a Border classic. Dr William Buchan's (1729–1805) *Domestic Medicine* is known all the world over. He was a native of Ancrum and was buried in Westminster Abbey. Dr William Turner was the surgeon who extracted the fatal bullet from Nelson's shoulder at Trafalgar. Sir Whitelaw Ainslie (1767–1837), medical writer, was born at Duns. Sir Andrew Smith (Heron Hall) (1797–1872) and Sir David James Dickson (Bedrule) (1780–1850) were notables in the medical world.

The *Minstrelsy of the Merse* (by the present writer) furnishes a list of 84 men and women who have written many well-remembered verses. Thomas of Ercildoune (? 1216–1294) has been styled the "day-star of Scottish poetry." Though Scott's claim for him as author of *Sir Tristrem* must be discarded in the light of recent research, Thomas's reputation rests upon other claims not so easily discounted. His reputed prophetic rhymes are still quoted, and his place in the fortunes of his country has an assured immortality. The Maitlands and Humes cut notable figures in poesy. Robert Crawford's *Cowdenknowes* and *Leader Haughs and Yarrow* are charming specimens of the native pastoral. Dr James Grainger (1723–1766), a Duns native, published a lengthy didactic poem on the *Culture of the*

Sugar Cane. Ralph Erskine's (1685–1752) *Gospel Sonnets* has been frequently reprinted. Anne Home (1742–1821)

Sir Walter Scott, Bart. *[Graham Gilbert*

wrote *My Mother bids me bind my Hair* and other lyrics, which Haydn wedded to inspiring music. Dr George Henderson of Chirnside (1800–1864), besides compiling the *Popular Rhymes of Berwickshire*, wrote many poems in praise of his native haunts. Thomas Telford, Andrew Wanless, Robert Maclean Calder, exiles in Canada, sang appealingly of old Merse scenes and memories.

John Leyden, M.D.

Roxburghshire possesses greater poetic names than the sister county. The spell of Sir Walter Scott (1771–1832), a Roxburghshire man by descent, by upbringing, and residence, hangs over the whole district. The *Lay of the Last*

Minstrel has almost its entire *locale* in Teviotdale. *Marmion* is associated with the shire. The scene of the *Eve of Saint John* was laid at Smailholm Tower. Many of the ballads of the *Minstrelsy* were "lifted" from the Liddesdale glens. Walter Scot of Satchells was the rhyming chronicler of the Scott clan. James Thomson (1700–1748), the poet of the *Seasons* and of *Rule Britannia*, was a native of Ednam. He spent his boyhood at Southdean. Dr John Leyden (1775–1811) was the most erudite Border man of his time. Born in a shepherd's cottage at Denholm, he rose to front rank as an Oriental scholar and Hindustani translator, with a knowledge of 27 languages and a passion for learning which has rarely if ever been excelled. His *Scenes of Infancy* is the classic poem of Teviotdale. He died in Batavia while serving with a British Expedition. Thomas Aird (1802–1876), journalist, and author of the weird *Devil's Dream*, was a native of Bowden. So was Andrew Scott (1757–1839), author of the whimsical *Rural Content*. Thomas Pringle was born at Blakelaw in Linton parish in 1789. He was one of the founders of *Blackwood's Magazine*. He became secretary of a Society whose aim was the emancipation of the South African slaves. As a poet his *Border Emigrant's Farewell* is often quoted. Much of his life was spent in Africa, but he died in London in 1834. His dust rests in Bunhill Fields not far from the grave of Bunyan. Henry Scott Riddell (1798–1870), author of the spirited *Scotland Yet*, was minister at Teviothead. Thomas Tod Stoddart (1810–1880), the angler-poet, belonged to Kelso. John Younger (1785–1860), another angling bard, piscatory authority and autobiographer, spent all his life at St Boswells as a working shoemaker. William Knox

(1789–1825), whose poem, *Oh, why should the spirit of mortal be proud*, was the favourite of President Lincoln and of a Czar of Russia, was a native of Lilliesleaf. James Telfer the balladist (1800–1862), was schoolmaster of Saughtree. Another who should be mentioned is Thomas Davidson, the "Scottish Probationer" (1838–1870), a native of Oxnam. Several hymn-writers are worthy of note. Dr Horatius Bonar (1808–1889), was a minister in Kelso. His hymns are the property of all the churches. Mary Lundie Duncan (1814–1840), writer of favourite children's hymns, was a daughter of the Kelso manse. Henry Francis Lyte (1793–1847), author of *Abide with Me*, was a native of Ednam. Elizabeth Cecilia Clephane (1830–1869), whose hymn *The Ninety and Nine* Ira D. Sankey sang into fame, had her home near Melrose. Anne Ross Cousin (1824–1906), authoress of *The Sands of Time are Sinking*, was the wife of a Melrose clergyman.

Localities so redolent of Nature's charms were bound to evoke the artistic temperament. Sir George Watson (1767–1837), first President of the Royal Scottish Academy, was born at Overmains, Eccles. His son, William Smellie Watson, R.S.A. (1796–1874), was an artist of repute. William Shiels, R.S.A. (1785–1857), came of Earlston stock. William Yellowlees (1796–1856), called "the Raeburn in little," belonged to Mellerstain. Robert Edmonstone, John Ballantyne, Alexander Hume and Edward Cunningham all belonged to Kelso. Andrew Currie (Darnick) (1812–1891) and Edwin Stirling (Dryburgh) (1819–1867) were well-known sculptors.

In music, there are fewer names. Thomas Legerwood Hately (1815–1867), composer of many fine psalm-tunes, belonged to Greenlaw. George Hogarth (1783–1870), author of a *History of Music*, hailed from Channelkirk. His daughter was the wife of Charles Dickens. John Thomson (1805–1841), Professor of Music at Edinburgh, was born at Sprouston. William Stenhouse (1773–1827), editor of Johnson's *Musical Museum*, was born near Bowden.

In the sphere of pure literature Sir Walter Scott is the commanding personality. Four of the "Waverleys" are connected with our counties. The *Bride of Lammermoor* is a striking instance of the transference of a Galloway tragedy to the south-east of Scotland. *Guy Mannering* abounds in picturesque descriptions of Liddesdale peasant life. In *The Monastery* and *The Abbot* Melrose—or "Kennaquhair"—and its monkish past find romantic immortality. At Chiefswood lived John Gibson Lockhart (1794–1854), Scott's son-in-law and biographer, who wrote some of his own novels there. Another occupant was Thomas Hamilton (1789–1842), who wrote at Chiefswood his *Cyril Thornton* and *Annals of the Peninsular Campaign*. G. P. R. James (1799–1860) the novelist, resided at Maxpoffle. Mrs Robert Logan's *St Johnstoun* and *Restalrig* had a wide vogue in their day. George Cupples (1822–1891), author of *The Green Hand*, was born at Legerwood. On the journalistic side, James Black (1783–1855), editor of the *Morning Chronicle*, was born in a ploughman's cottage at Burnhouses, near Duns. James Cleghorn (1778–1838) from Duns, was first editor of *Blackwood's Magazine*. The brothers Ballantyne—James, John, Alexander, who so potently influenced the fortunes, or misfortunes, of Scott, hailed from Kelso. Alexander was father of R. M.

Ballantyne, favourite of boys all the world over. James George Edgar (1834–1864), from Hutton, was another prolific writer of boys' stories. Andrew Wilson (1831–1881), son of the Indian missionary, edited the *Times of India* and wrote *The Abode of Snow*. William Jerdan (Kelso) (1782–1869), the critic, came into prominence from having been in conversation with Mr Percival in the lobby of the House of Commons when Bellingham fired his deadly shot at the Prime Minister. At great risk Jerdan seized the assassin and handed him over to justice. In James Sibbald's (1745–1803) *Edinburgh Magazine* appeared the first serious review of Burns's Poems in 1786. Sibbald's *Chronicle of Scottish Poetry* is a well-known work. He was born at Whitlaw in Roxburghshire. Sir William Robertson Nicoll (1851–1923), one of the greatest of modern journalists, was a Kelso minister. Anna M. Stoddart (1840–1911) of Kelso wrote the life of John Stuart Blackie, himself come of Kelso forebears. Mary Monica Maxwell Scott (1852–1920), great-granddaughter of Sir Walter Scott, wrote several Memoirs. William Scott Douglas (1815–1883) was a leading Burns authority. William Brockie (1811–1890), a native of Lauder, wrote *The Gypsies of Yetholm* and other Border books.

Religious writers were Robert Ainslie (1766–1838), a native of Duns, James Douglas of Cavers, and Alexander Hislop (Duns), author of the *Two Babylons*. Distinguished philosophical writers were Dr Andrew Wilson (1718–1792) and Professor David George Ritchie (1853–1903), born at Maxton and Jedburgh manses respectively. Dr Thomas Kirkup (1844–1912) from Yetholm, was the historian of Socialism.

Great publishers were James Nisbet, London, born at Spylaw, Kelso, and the brothers Carter, from Earlston, founders of what in its day was the most famous book-store in New York.

Scholarship has notable representatives within our counties. Many adorned the Professoriates. Classical pundits were George Dunbar (1774–1851), who began life as a gardener's boy at Ayton, learned to read Greek, and occupied that Chair at Edinburgh. Dr William Veitch's (1794–1885) *Greek Verbs* gained him a European reputation. He was born at Spital-on-Rule. Alexander Christison (1749–1820) was a renowned teacher of Latin. Abraham Robertson (1751–1826) was a pedlar in early youth. From Duns he found his way to Oxford and became Professor of astronomy. Robert Blair (died 1828) taught astronomy at Edinburgh. He was inventor of the 'aplanatic' telescope. Prominent medical teachers were Sir Robert Christison (1797–1882), Robert Lee (Melrose) (1793–1877) and William Rutherford (Ancrum) (1839–1899). Robert Hislop (Duns) and Robert Trotter were educationists of front rank. Sir James Augustus Henry Murray, F.B.A., LL.D., D.C.L., D.Litt., Ph.D. (1837–1915), greatest of British lexicographers, was a native of Denholm. He began life as teacher in a Hawick school and in 1858 was Master of Hawick Academy. From 1870 to 1885 he was Master at Mill Hill School, London. He read assiduously every book that came to his hand, unconsciously preparing himself for the great work of his life, the unparalleled *New English Dictionary on Historical Principles*.

We should be proud to call the profound philosopher

John Duns Scotus a Borderer. His epitaph in the Cathedral
of Cologne, where he died in 1308, begins *Scotia me genuit*;
but *Scotia* here may not mean our country. Only am-
biguous evidence exists for the birthplace of this great

Sir J. A. H. Murray

schoolman, the "Subtle Doctor," renowned for the depth
and breadth of his intellectual penetration. Places in Ireland
and England as well as Duns claim him as a native. Though
born in Edinburgh, David Hume (1711–1776), the greatest
of Scottish philosophers, was the son of a Berwickshire laird.
Ninewells was his home and there he spent his boyhood.

Distinguished naturalists include John (1799–1861), and William Baird (1803–1872), natives of Swinton, Dr James Hardy (1815–1898), the indefatigable historian of the Berwickshire Naturalists' Club, Dr Francis Douglas, James Duncan, a noted entomologist, and Dr Thomas Jerdon (1811–1872), a noted zoologist. Dr Robert Hogg (1818–1897) from Duns, founded the British Pomological Society. George Sinclair (1786–1834) from Mellerstain, and Robert Fortune (1813–1890) from Edrom, were botanists of distinction. The latter was superintendent at Kew Gardens. His *Three Years Wanderings in China* is a fascinating record.

Among scientists must be mentioned the names of James Bassantin (died 1568). He became Professor of Mathematics at Paris, but retired to Scotland where the remainder of his life was spent on his Bassendean estate. Dr James Hutton (1726–1797), author of the *Theory of the Earth*, was laird of Slighhouses, Bunkle. Sir David Brewster (1781–1868), son of the schoolmaster of Jedburgh, was the master of optics in his day. He invented the kaleidoscope and was Principal of Edinburgh University. Many of his instruments were fashioned in the workshop of his friend James Veitch (1771–1838), the self-taught astronomer of Inchbonny. Mary Somerville (1780–1872), one of the greatest of women scientists, was born in the manse of Jedburgh. Sir Peter Fairbairn (1789–1874) and Sir James Brunlees (1816–1892), eminent engineers, belonged to Kelso. George John Romanes (1848–1894), the biologist, was of Berwickshire extraction. James Lee Paris invented the rifle called by his name. Andrew Rodger invented the winnowing-machine.

A number of historians belong to our shires. Dr Thomas Somerville (1741–1830) occupied himself with Queen Anne's reign. George Ridpath (1717–1772) compiled an elaborate *Border History*. Dr George Barry (1748–1805) was historian of the Orkney Islands. Local historians are abundant: James Morton's *Monastic Annals of Teviotdale* narrates the story of the Border Abbeys. Robert Bruce Armstrong published a masterly account of Liddesdale; James Watson told the story of Jedburgh Abbey; David Gilmour Manuel, that of Dryburgh. Jane Rutherford Oliver produced an exhaustive record of *Upper Teviotdale and the Scotts of Buccleuch*. James Wilson wrote the history of Hawick, James Wade that of Melrose, James Haig that of Kelso. George Tancred of Weens amassed a pile of genealogical information in *Rule Water and its People*. Riddell Carre's *Border Memories* deals with the notable families of Roxburghshire. James Tait's *Border Church Life* is a painstaking record.

Men of action—naval and military heroes—in addition to those already mentioned, comprise Sir Thomas Brisbane *Makdougall (1773–1860)*, of Makerstoun, who fought through the Peninsular campaign and became Governor of New South Wales. The capital of that colony was named after him. He excelled as an astronomer. Admiral Sir James Douglas (1703–1787) of Springwood Park, brought to Britain the news of the storming of Quebec by Wolfe in 1759. He took Dominica, and was present at the siege of Martinique. His son, Admiral James (1755–1839) had a distinguished naval career. From Sydenham sprang Admiral Sir Archibald Collingwood Dickson (1772–1827), Admiral Sir William Dickson (1798–1868) and General Sir Alexander Dickson (1777–1840). Other notable sailors were Sir George Elliot, Robert Elliot, George Scott, Lord Mark Kerr, Sir William and Sir Henry Fairfax, and Thomas Baillie—Admirals all. Other great soldiers were Andrew Rutherford, Earl of Teviot, and Sir Alexander James Hardy Elliot.

Of travellers abroad and men whose careers were spent in far-away lands, the following may be noted: Patrick Brydone (1736–1818), born at Coldingham, was author of *A Tour through Sicily and Malta*, in its day a widely read book. He lived at Lennel and was the "pious pilgrim" mentioned in *Marmion*. James Brown (1709–1788) sojourned in Russia and Persia. He compiled a Persian dictionary and grammar. Dr Alexander Anderson (1770–1805), Mungo Park's brother-in-law, who perished on the last Niger expedition, was born at Earlston. John Parish Robertson (Kelso) (1792–1843) and his brother William laid the fortunes of the Argentine. The brothers Chirnside were Australian wool kings. John Redpath (1796–1869), from Earlston, was a pioneer of modern Montreal. His son, Peter Redpath (1821–1894), was a lavish benefactor of McGill College. James Bell (1769–1833) was a well-known geographer. Andrew Elliot was Governor of New York. Alexander Spottiswood (1676–1740) was Governor of Virginia. There is good ground for believing that the grandparents of Captain James Cook hailed from Ednam.

Successful merchants were Sir Peter Laurie (1778–1861), son of a Stichill farmer, and Sir John Pirie (1781–1851), a native of Duns. Both became Lord Mayor of London. Sir John Marjoribanks and Sir John Boyd were Lord

Provosts of Edinburgh. William Mills was Lord Provost of Glasgow. William Jacks (1841–1907), son of a Merse shepherd, was iron-master, Member of Parliament, linguist, biographer of Bismarck and James Watt. Sir Robert Laidlaw (1856–1917) was an India merchant. He was born at Bonchester. Politicians include David Robertson of Ladykirk, created Baron Marjoribanks (1797–1873), Arthur Ralph Douglas Elliot (1842–1924), and Edward Marjoribanks, second Lord Tweedmouth (1849–1909). Among Indian statesmen were Sir Daniel Eliott (1798–1872), James Wilson of Hawick (1805–1860), and Sir Henry Maine (1822–1888). Robert Ramsay (Hawick) (1842–1882) was an Australian statesman.

From the Border counties it was to be expected that Scott would borrow many of the Originals of his great characters. The greatest of these he found in his own country. Robert Paterson of Hawick was prototype of "Old Mortality"; the gaberlunzie, Andrew Gemmels, stood for the portrait of "Edie Ochiltree"; Willie Elliot of Millburn for that of "Dandie Dinmont"; George Thomson (Melrose) and James Sanson (Earlston) comprised the picture of "Dominie Sampson." Dr Duncan of Smailholm was sketched in the "Rev. Josiah Cargill"; Jean Gordon, Yetholm gypsy, was the real "Meg Merrilies."

To end on a royal note, Alexander III—"whom Scotland led in luve and lee"—was born at Roxburgh Castle in 1241.

23. The Chief Towns and Villages.

[The figures in brackets after each name give the population in 1921, and those at the end of each section are references to the pages in the text.]

BERWICKSHIRE.

Abbey St Bathans (203), a parish and village in Lammermoor, five miles south-west of Grantshouse. Called Abbey St Bathans to distinguish it from the better known St Bathans at Yester. (pp. 4, 10, 22, 153.)

Ayton (1521), a coast parish and village of the same name containing also the fishing village of Burnmouth (802). The parish church, a fine First Pointed edifice, was erected in 1865. (pp. 22, 52, 65, 79, 81, 137, 149, 150, 152, 167, 176.)

Bemersyde, a hamlet in Mertoun parish. Of Bemersyde House Thomas the Rhymer (? 1290) said:

"Tyde, tyde, what may betyde,
Haig shall be Haig of Bemersyde."

(pp. 20, 49, 82, 88, 124, 128, 134, 161.)

Chirnside (1402), a parish and village in the east of the county, six miles from Duns. Situated on the brow of a hill, the prospect is wide and commanding. Chirnside Church, a venerable structure, was extensively restored in 1907. Henry Erskine was minister from 1690 to 1696, and Dr Walter Anderson, author of a *History of France*, from 1756 to 1800. (pp. 5, 79, 102, 124, 137, 150, 155, 170.)

Cockburnspath (941), abbreviated into Co'Path (anciently Coldsbrandspeth), sometimes vernacularly "Copperspath," a maritime parish and village in the north of the county. Its Cross is one of the oldest in Scotland. (pp. 10, 44, 98, 102, 106, 118, 124.)

Coldingham (2830), a coast parish and village, and one of the oldest places in the shire. Anciently known as Coldinghamshire, it lay ecclesiastically within the diocese of Lindisfarne. Coldingham Moor, extending to about 6000 acres has been considerably reclaimed and brought under the plough. Auchencraw and Grantshouse (485) are villages in the parish. (pp. 10, 27, 32, 40, 50, 55, 65, 67, 106, 117, 166, 180.)

Coldstream (2013), anciently Lennel, a parish and town in the south of the shire, finely situated on the left bank of the Tweed and its tributary, the Leet. Cornhill 1½ miles off, on the English side, is the station for Coldstream. It is reached by Coldstream Bridge, built in 1763–1766 by Smeaton. At its Scottish end stands the cottage famous as a venue of runaway weddings until the Marriage Act of 1856 put an end to them. Lord Brougham's marriage took place in an *inn* of the town, not at the Bridge, as is generally stated. In earlier

Coldstream Bridge

days a ford at Coldstream was the only means of passage from north or south. By it Edward I invaded Scotland in 1296, and it was used by numerous armies and marauding bands. (pp. 21, 23, 34, 66, 103, 137, 145, 152.)

Cove, a fishing hamlet in Cockburnspath parish. Its harbour is approached by a rock-hewn tunnel, 65 yards long, wide enough to admit a horse and cart. The coast scenery is picturesque. (pp. 50, 57, 86.)

Duns (2818), spelled Dunse from 1740 to 1882, the original Duns being then restored, a parish and county town of Berwickshire. The town lies on the skirts of Duns Law, at 437 feet above sea-level. An earlier town was destroyed by the English in 1545. An object of reverential interest on the summit of the Law—a round, turf-clad hill—is the "Covenanters' Stone," marking the position occupied by Leslie's army in 1639. Modern Duns dates from 1588. From 1661 to 1696 it was the county town, the honour then going to Greenlaw. In 1882 Duns recovered that distinction. (pp. 28, 48, 67, 75, 106, 135, 143, 166, 167, 168, 176.)

Earlston (1643), anciently Ercildoune (*arciol-dun*, the look-out hill), a parish and small town on the Leader close to the Roxburgh-shire border. David I signed "apud Ercheldon" the foundation Charter of Melrose Abbey in June 1136, and Prince Henry subscribed here its confirmatory Charter in 1143. On his way to Flodden, James IV "campit ane nicht in Ersilton." Prince Charlie passed through in 1745. The Crawford Lindsays were founders of the place. After them came the Earls of Dunbar. The Earl of Haddington is now "Superior." Thomas the Rhymer's Tower is the chief object of antiquity. A stone in the parish church is inscribed:

> "Auld Rymr Race
> Lyes in this place."

Redpath (67) and Mellerstain are villages in the parish. Fans, a decayed hamlet, now a farm-place, belonged to the Gordons. (pp. 9, 10, 17, 22, 27, 40, 65, 75, 78, 79, 96, 102, 106, 126, 132, 161, 167, 169, 176, 180.)

Eccles (1245), a parish and village bordering on Roxburghshire. It contains also the villages of Leitholm (607) and Birgham. (pp. 11, 21, 88, 106, 149, 168, 173.)

Edrom (1119), a parish and village on the Whitadder in the east centre of the Merse, about three miles from Duns. It contains the village of Allanton. At "Battie's Bog," the Chevalier de la Bastie, a French knight to whom the Regent Albany had committed the Wardenship of the Marches, was atrociously slain by Home of Wed-derburn in 1517. (pp. 22, 65, 79, 98, 150, 178.)

Eyemouth (2573) is eight miles north-west of Berwick. A small but substantial harbour is protected by the "Harkers" and a break-water. A title of the great Duke of Marlborough was Baron Eyemouth in the Scottish peerage. (pp. 10, 22, 32, 41, 45, 50, 52, 80, 84, 146, 150.)

Fogo (383), a central parish and hamlet, 4½ miles from Duns. Fogo Church, a simple, picturesque edifice, overlooks the Blackadder. A curious memorial panel in the outside wall bearing the figures of two men in broad-skirted coats and full-bottomed wigs, and of a woman in a robe with a sash or girdle round her waist carries the legend:

"We three served God: lived in His fear,
And loved Him who bought us dear."

(p. 22.)

Foulden (253), a parish and village in the eastern part of the Merse. In Foulden Church on 24th March, 1587, Queen Elizabeth's commissioners met with those of James VI to discuss the execution of Queen Mary. (pp. 61, 73, 99.)

Gavinton, a village in Langton parish (385), built in 1760 by the Marquess of Breadalbane, and named in honour of David Gavin, his maternal grandfather. Langton House (p. 137) was demolished in 1925.

Gordon (719) a parish and village in the west of the Merse. Greenknowe Tower was the abode of Walter Pringle the Covenanter. (pp. 27, 36, 41, 44, 150, 161, 167.)

Greenlaw (909), a parish and small town on the Blackadder The village originally stood about a mile and a half to the south-east on an isolated green "law" or hill, from which it takes its name. Modern Greenlaw dates from the close of the seventeenth century. From 1696 to 1882 it was the county town. (pp. 32, 33, 106, 145, 150, 167, 174.)

Houndwood (1188), a village on the left bank of Eye Water, and from 1851 a *quoad sacra* parish disjoined from Coldingham. (p. 124.)

Hume (258), a parish conjoined, for ecclesiastical purposes only, with Stichill in Roxburghshire. Each parish has its own school, and its own registration officials. The tiny hamlet nestles at the base of the rocky eminence crowned by Hume Castle. In its old church-yard is a mound called the "Pest Knowe"—the burial-place of victims of the Plague in 1645. (pp. 10, 27, 122.)

Hutton (655), a parish and village containing also the hamlet of Paxton, between the Whitadder and the Tweed. Near Paxton is the Union Suspension Bridge 361 feet in length, built in 1821 by Sir Samuel Brown—the first of its kind to be erected in this country. (pp. 124, 137, 165, 175.)

Ladykirk (338), anciently Upsetlington, a parish and village on the north bank of the Tweed facing Norham Castle. The church was a constant meeting-place for the Wardens of the Marches, and the last Treaty between Scotland and England was signed within its walls. It was restored in 1861. Horndean is a village in the parish. (p. 117.)

Lauder (1369), a parish and only royal burgh of Berwickshire, is situated in the north-west of the county, 25 miles from Edinburgh. A place of great antiquity, it figures frequently in Scottish history. The church in which "Bell-the-Cat" and his confederates met to plan the removal of James III's favourites stood within Thirlestane grounds, and has disappeared. (pp. 11, 22, 44, 76, 78, 92, 98, 145, 152, 168, 175.)

Legerwood (379), a Lauderdale parish and hamlet four miles from Earlston. The chancel of its ancient church (restored in 1898) contains the grave of Grisell Cochrane, wife of John Ker of Morriston, who saved the life of her father, Sir John Cochrane of Ochiltree, when under sentence of death at Edinburgh for political offences in 1685. This she did by robbing the mail-bag which contained his death-warrant. William Calderwood, the Covenanter, was minister here 1655–1709. (pp. 27, 98, 126, 174.)

Longformacus (250), a parish and village on the Dye Water in Lammermoor, seven miles from Duns. At Longformacus manse was born Thomas Ord (1785–1859), the most noted equestrian of his time. (pp. 4, 10, 81, 98.)

Mertoun (463), a parish and hamlet in the extreme south of the shire. The Tweed forms picturesque loops round Dryburgh Abbey and Bemersyde. Sir Walter Scott's favourite view overlooked the latter. There are hamlets at Dryburgh, Clintmains, and Bemersyde. A colossal statue of Wallace, the first to be erected in Scotland—in 1814—is on a wooded eminence near Bemersyde. (pp. 9, 10, 88, 134, 168.)

Mordington (299), a coast parish and hamlet in the extreme south-east of the shire, four miles from Berwick-upon-Tweed, containing also the small fishing village of Ross. Bernard de Linton, Chancellor of Scotland, was rector in 1296. (pp. 49, 81, 126.)

Nenthorn (363) (anciently Nattaisthyrne, probably Nechtan's Thorn), a parish and village in south-west Berwickshire, and about four miles from Kelso. (pp. 10, 27, 41, 165, 168.)

Oxton (188), anciently Ulfkiliston, a village in Channelkirk (anciently Childeschirche) parish (499), five miles from Lauder, with a station on the light railway from Fountainhall, is a favourite summer resort. (pp. 65, 150.)

Polwarth (167), a parish and village in the centre of the Merse. Polwarth Church dates from 900 A.D. Under its east end is the

Dryburgh Abbey

vaulted burial-aisle in which Sir Patrick Hume was concealed in 1684. A pulpit frontal, still extant, was the work of his daughter, Lady Grisell Baillie. (p. 126.)

Reston (1271), a village in Coldingham parish. (pp. 32, 75, 150.)

St Abbs (403), formerly Coldingham Shore, a fishing village and summer resort in Coldingham parish. (pp. 35, 52, 55, 62, 86.)

Swinton (689), a parish and village in the eastern part of the Merse. Its church bell (1499) is the oldest in the Border. It is inscribed *Maria est nomen meum.* (pp. 23, 40, 82, 118, 137, 178.)

Westruther (448), anciently Wolfstruther, a Lammermoor parish and village six miles from Gordon, containing the decayed hamlet of Houndslow. The Twinlaw Cairns (1466 feet) are associated

Polwarth Church

with a tradition in which two Edgars of Wedderlie, ignorant of their identity, were slain in a duel. (pp. 4, 27, 36, 40, 99, 126, 133, 168.)

Whitsome (473), a parish and village in the Merse, entirely devoted to agriculture.

ROXBURGHSHIRE.

Ancrum (912), anciently Alnecramb, a parish and village on the right bank of the Ale, 3½ miles north-west of Jedburgh. The Mantel (not, as often given, Malton) Walls, were remains of outworks of a fort or tower for the defence of Ancrum. They have no connection with the Knights of Malta. (pp. 25, 48, 105, 106, 117, 149, 167, 169, 176.)

Bedrule (180), a central parish and hamlet on Rule Water, about two miles east of Denholm. Its long-demolished castle was a stronghold of the Turnbulls. The Dunian view is one of the finest in the Border. (pp. 67, 127, 169.)

Bonchester Bridge, a hamlet in Hobkirk parish (555), eight miles from Jedburgh. Bonchester Hill (1059 feet) has remains of ancient fortifications, thought to have been occupied by the Romans under the name of "Bona Castra." (pp. 26, 97, 181.)

Bowden (638), anciently Bothenden, a parish and picturesque village in the north-west of the county, about two miles from Newtown St Boswells station. Its ancient Cross has been restored as a War Memorial. The parish contains also the small village of Midlem (98). (pp. 25, 96, 106, 120, 172.)

Crailing (488), a parish and village on Oxnam Water, containing also the hamlet of Nisbet. (pp. 26, 105, 167.)

Darnick, a village in Melrose parish. Its chief feature is Darnick Tower. (pp. 20, 129, 173.)

Denholm (384), a village in Cavers parish (582), two miles east of Hassendean, and five from Hawick. The birth-cottage of John Leyden, the property of the Edinburgh Border Counties Association, contains many relics—letters, manuscripts, portraits—of the scholar and poet. His monument stands on the village green. (pp. 24, 45, 65, 172, 176.)

Eckford (665), a parish and village on the right bank of the Teviot, near the influx of the Kale, 1½ miles from Kirkbank station. (pp. 26, 99, 168.)

Edgerston (229), from 1855 a *quoad sacra* parish, with a village about eight miles from Jedburgh. (pp. 26, 32, 48, 97, 155.)

Ednam (413), a parish and village on the Eden, two miles from Kelso. King Edgar (1097–1107) granted lands now forming this parish to Thor, a Northumbrian. Thor built a church dedicated to St Cuthbert, and endowed it with a ploughgate of land—104 acres. This is the earliest record of the erection of a parish in Scotland. On the Ferney Hill, the Ednam Club erected in 1820 an obelisk to the memory of James Thomson. (pp. 27, 48, 65, 80, 140, 172, 173, 180.)

Gattonside (272), a beautiful village on the north bank of the Tweed in Melrose parish. Many places bear old monastic names which gave it origin—the Abbot's Meadow, the Vineyard, Friar's Close, and Cellary Meadow. Its orchards planted by the monks have long been famous for the quality and quantity of their fruit. (pp. 98, 103.)

Hawick (17,445), *haga-wic*, "the fenced-in dwelling," a parish and large manufacturing town in Teviotdale. Though one of the oldest places in the county almost all marks of antiquity have vanished. These consist now chiefly of The Moat, St Mary's Church and the Baron's Tower (a hotel), built originally by the Lovels of Branxholme and Hawick. The great living link between Hawick's past and present is the festival of the Common Riding. There is more than an official formality in this annual "riding of the marches." It is a patriotic demonstration—a perpetual reminder of the incident mentioned at p. 92. No Border town has become so extended or has so altered its appearance. In 1778 Hawick was a place of small importance. It had no post-office, and its letters were brought by a common hawker from Jedburgh. In 1801 the population was 2798. By 1861 it had risen to 7801, and in 1891 it was 14,122. It is now the most important town in the district, consisting of the two parishes of Hawick and Wilton (3436), with three *quoad sacra* parishes. It is the principal centre of the Scottish woollen industry. (pp. 15, 23, 25, 35, 41, 44, 67, 75, 78, 92, 99, 146, 166.)

Hownam (191), a Cheviot hill parish in the east of the shire, with a village of the same name on the right bank of the Kale, 11 miles from Kelso. (pp. 26, 103.)

Jedburgh (3533), originally Jedworth, a parish and the only royal burgh of the county. As the chief town on the East Border it was the scene of stirring conflicts and notable assemblies. Its male population were known as valorous fighters with their death-dealing axe or "staff"—a stout steel-headed pole four feet long, and by their terror-striking slogan, "Jethart's here." Surrey declared that in his storming of the town he found the men of Jedburgh "the boldest

and the hottest that ever I saw of any nation." From Bannockburn
to Killiecrankie they played a conspicuous part in the Scottish
struggle. "Jeddart Justice," like the Devonshire "Lydford Law"—

> "How in the morn they hang and draw,
> And sit in judgment after,"

was an old opprobrium having no foundation in fact. The original
Jedworth, founded by Bishop Ecgred of Lindisfarne, stood a few
miles further up to Jed. (pp. 15, 26, 49, 76, 79, 91, 92, 110, 146, 153,
175, 178.)

Kelso (4009), a parish and market town in the richest part of
Roxburghshire at the meeting of Tweed and Teviot. Kelso station
is in the village of Maxwellheugh, a mile off. Sir Walter Scott spent
part of his boyhood with his aunt at Garden Cottage, now Waverley
Lodge. His great-grandfather "Beardie," and many of his ancestors,
lie buried close under the Abbey walls. (pp. 21, 23, 40, 65, 75, 89,
112, 146, 174, 178.)

Lilliesleaf (549), anciently Lillesclive, a parish and pretty
village in the north-west of the county, three miles from Belses
station. The church, mentioned in the "Inquisition of David" in
1116, was finely restored in 1910. Lilliesleaf Moor was a conventicle
centre. (pp. 25, 40, 78, 155, 173.)

Linton (410), a rural parish and hamlet in the north-east of the
county, bordering on Northumberland and about 6¼ miles from
Kelso. (pp. 26, 28, 40, 99, 119, 172.)

Makerstoun (277), anciently Malkariston, a small parish and
village on the northern boundary of the county. Makerstoun House
occupies a prominent site amidst stately woodlands. Trow Crags on
the Tweed are an interesting feature. (pp. 21, 103, 179.)

Maxton (357), anciently Maccus-toun, a parish and village on
the north border of the county and right bank of the Tweed with
a railway station on the St Boswells—Kelso line. Its ancient Cross
was restored in 1882. (pp. 34, 40, 106, 120, 150, 155, 175.)

Melrose (4536), a parish and picturesque town on the Tweed,
designated the "capital of the Scott Country." Abbotsford is within
3 miles, and Dryburgh 5½ miles distant. The original Melrose stood
3 miles further down the river at the spot called Old Melrose. There,
about 650, was founded a primitive abbey, at first a cell of St
Aidan's monastery of Lindisfarne. No remains are extant of this

ancient shrine. Several place-names—the Monk's Ford, the Holy Wheel (i.e. pool), the Chapel Knowe, and the Girth-Gate preserve the memory of its former sanctity. In 1136, David I founded the new Melrose, its site being then called Fordel. Melrose Abbey lives more through the genius of Sir Walter Scott than even by its historical associations. Many notable people lie buried within its walls. Blainslie and Langshaw are hamlets in the parish. (pp. 10, 20, 65, 76, 92, 106, 115, 173, 174.)

Minto (382), a parish and village on the left bank of the Teviot equidistant five miles from Hawick and Jedburgh. Fatlips Castle (a reconstruction) is romantically placed on the eastern slope of Minto Crags. The ruins of Barnhills Castle are in a glen to the east of the Crags. (pp. 48, 142, 163, 165.)

Morebattle (774), anciently Merboda, Merebottle, a parish and village in the east of the county. The village stands on a gentle eminence 320 feet above sea level near the left bank of the Kale 4½ miles from Yetholm, and 7½ from Kelso. The entire parish consists of hills and narrow valleys running up to the highest summits of the Cheviots and forming a direct boundary with Northumberland. Corbet Tower was burned by Hertford in 1545. Though rebuilt 30 years later it again fell into decay and was partly renovated by a nineteenth century proprietor. Whitton Tower was also burned in 1545 with other peels in the parish. (pp. 26, 81, 99.)

Newcastleton (867), a small town in Castleton parish (1853), Liddesdale, so named in contradistinction to the decayed village of Castleton, which stood a little further up the valley. (pp. 75, 129, 150.)

Newstead (208), a village of considerable antiquity in Melrose parish. The important Roman station of Trimontium was close by. (pp. 20, 82, 104.)

Newtown St Boswells (592), a village and railway centre in Melrose and St Boswells parishes. Stock sales are regularly held here. (pp. 75, 154.)

Oxnam (550), a parish and hamlet on Oxnam Water, 4½ miles south-east of Jedburgh. Dolphinston and Henwood were fortalices of strength and note. The latter gave origin to the memorable war-cry "A Henwoody!"—to rouse and lead a Border onset. (pp. 15, 26, 173.)

Roberton (383), a parish on the Borthwick Water, containing the hamlet of Deanburnhaugh about eight miles from Hawick. A well-defined portion of the Catrail runs through the district. (pp. 13, 14, 25.)

Roxburgh (742), a parish and village on the south side of the Tweed, three miles from Kelso. The Teviot intersects it from north to south. Heiton is a village on the south side of that river. Roxburgh Castle is in the parish. Of Old Roxburgh, one of the first four royal burghs in Scotland, not a trace remains. It stood on the level sward immediately in front of the ancient castle ruin, where St James's Fair is held. In Roxburgh churchyard was buried, in 1798, Andrew Gemmels, aged 106, the Original of "Edie Ochiltree." A sculptured stone marks his grave. (pp. 2, 21, 90, 127, 148, 150, 181.)

St Boswells (950), a parish and village on the south bank of the Tweed opposite Dryburgh. Anciently called Lessudden, from St Aidan, it is a place of considerable antiquity and contained no fewer than 16 peels and bastle-houses. All were destroyed by the English in 1544. St Boswells is headquarters for the Duke of Buccleuch's Hunt. (pp. 75, 120, 148, 150, 172.)

Smailholm (261), a parish with a long straggling village or group of villages, six miles from Earlston, and six miles from Kelso. At Sandyknowe farm Sir Walter Scott resided with his grandparents from his third to his sixth year. Smailholm Tower, close by, inspired many fine lines in *Marmion*. Wrangholm was the Rhuringham of Bede's *Life of St Cuthbert*. (pp. 17, 27, 34, 65, 119, 128, 172, 181.)

Southdean (533), anciently Chesters, a parish of south-east Roxburghshire, containing the hamlet of Chesters, near the left bank of the Jed, seven miles from Jedburgh and about nine miles from Hawick. Close to the English border is the locale of the "Raid of the Redeswire," where on 7th July, 1545, a number of Scots, resenting the slaughter of one of their countrymen, attacked the offenders and were repulsed. In retreating they fell in with a band of Jedburgh men, who joined them, and wheeling back, completely routed their pursuers. James Thomson's father was minister of the parish 1700–1716, and here the poet spent his boyhood. A new church, erected in 1876, contains a window to his memory. (pp. 15, 92, 98, 99.)

Sprouston (776), a Tweedside parish abutting on the English border, and containing a village of the same name with the hamlet of Lempitlaw, a little over two miles from Kelso. At Hadden Stank and Redden Burn commissioners for settling Border disputes frequently met. Hadden Rig (541 feet) was the scene of a skirmish about 1540, when 3000 English horsemen were put to rout by a small body of Scottish troops. (pp. 21, 32, 65, 82, 151, 174.)

Stichill (274), a parish with a village 405 feet above sea level in the north of the shire and 3¼ miles from Kelso. Stichill House, built in 1866 on the site of an earlier structure, is a handsome edifice with a tower 100 feet high commanding magnificent panoramas for 30 miles round. The policies possess much beauty. The estate has belonged in turn to the Lochinvar Gordons, the Pringles, and the Bairds of Gartsherrie, the last of whom was George "Abington" Baird, the gentleman jockey. Stichill Linn is a high waterfall on the Eden. (pp. 11, 27, 34, 169, 180.)

Teviothead (377), anciently Carlanrig, a parish and hamlet near the source of the Teviot. Henry Scott Riddell's monument is a conspicuous object on the Dryden Knowes. (pp. 149, 172.)

Yetholm (772), a hill parish and village amongst the Cheviots, close to the English border, and 7½ miles from Kelso. It consists of two parts—Town and Kirk Yetholm—lying on each side of the Bowmont. Kirk Yetholm was long the headquarters of Scottish gypsydom. The last king, Charles Faa Blythe, crowned in 1889, died in 1902. (pp. 14, 28, 65, 78, 99, 106, 167, 175, 181.)

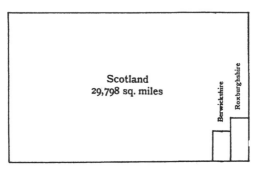

Fig 1. Areas (excluding water) of Berwickshire (457 sq. miles) and Roxburghshire (666 sq. miles) compared with that of Scotland

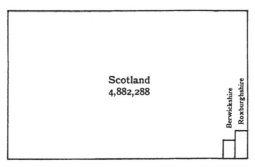

Fig 2. Population of Berwickshire (28,246) and of Roxburgh-shire (44,989) compared with that of Scotland in 1921

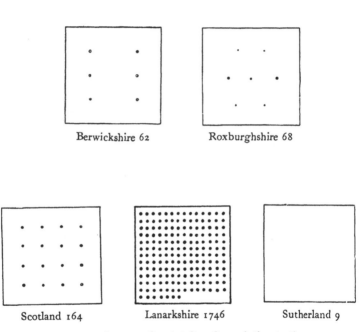

Berwickshire 62   Roxburghshire 68

Scotland 164   Lanarkshire 1746   Sutherland 9

Fig. 3. Comparative density of population to the
square mile in 1921
(*Each dot represents* 10 *persons*)

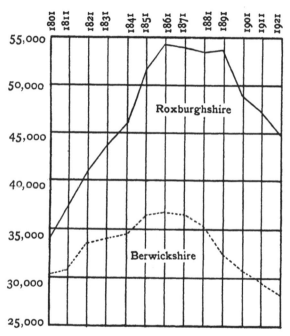

Fig. 4. Graph showing rise and fall of population in Berwickshire and Roxburghshire, 1801–1921

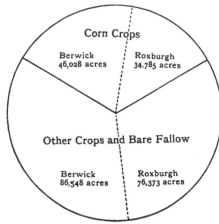

Fig. 5. Proportionate areas under corn crops compared with
that of other cultivated land in Berwickshire and Roxburghshire in 1924.

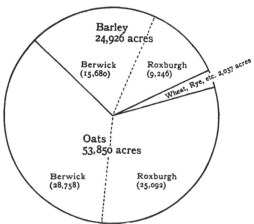

Fig 6. Proportionate areas of chief cereals in Berwickshire
and Roxburghshire in 1924.

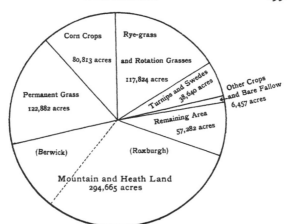

Fig 7. Proportionate areas of land in Berwickshire
and Roxburghshire in 1924

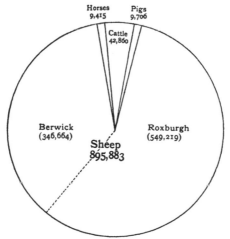

Fig. 8. Proportionate numbers of live stock in Berwickshire
and Roxburghshire in 1924

Milton Keynes UK
Ingram Content Group UK Ltd.
UKHW032321161024
449665UK00001B/7